MW00824202

Excellent Joy

EXCELLENT JOY

Fishing, Farming, Hunting and Psychology

Michael R. Rosmann

Ice Cube Books
North Liberty, Iowa

Excellent Joy:
Fishing, Farming, Hunting, and Psychology.

Library of Congress Control Number: 2011920585

ISBN 9781888160550

Ice Cube Books, LLC (est. 1993)
205 North Front Street
North Liberty, Iowa 52317-9302
www.icecubepress.com
steve@icecubepress.com
1-319-558-7609/f 319-626-2055

Manufactured in USA (2018 digital edition)

The paper used in this publication meets the minimum requirements of the American National Standard for Information Sciences—Permanence of Paper for Printed Library Materials, ANSI Z39.48-1992

"Idaho Crappie Fishing In Winter": an earlier version of this story was published in *North American Fisherman*, Vol. 15, No. 7 (2002). "The Meaning of Christmas" was originally published in *The Des Moines Register*, January 6, 2008. "The Behavior of Farm People" was originally published in the *Wapsipinicon Almanac*, No. 14 (2007), Anamosa, IA.

Cover photo courtesy of Steve Kowalski, A Better Exposure.
All line drawings by Michael R. Rosmann

Table of Contents

Foreword

In the May/June 2008 issue of *Orion* magazine, Jeffrey Kaplan published an essay entitled "The Gospel of Consumption." In the article Kaplan points out that during the early 1900s Americans began a major transition from being a rather frugal society to one that gradually became driven by consumption. But, this shift was not sufficient to satisfy our emerging industrial economy. Despite the increasing levels of consumption "industrialists were worried" because they knew the capacity for our industrial economy to produce goods exceeded "people's sense that they needed them." To correct this potential imbalance our industrial economy developed a "strategic shift for American industry—from fulfilling basic human needs to creating new ones." This strategy created a culture in Western societies which convinced most of us that we need much more material wealth than we actually need—that our well-being is somehow inextricably tied to the wealth of goods that we can accumulate. This culture is driven by a "gospel of consumption."

The evolution of this culture among us led to at least two unintended consequences. First, it led us to conclude that the only way we can have a healthy economy is through unlimited growth, a concept, that Herman Daly reminds us, is now so fixed in our society that any effort to challenge it is one of our culture's great "anathemas." Furthermore, the unintended consequence of this non-negotiable principle in our society is the ruination of our ecological capital, a fact that now puts the lives of our children and grandchildren in jeopardy. A second unintended consequence of the "gospel of consumption" is that it has actually deteriorated, rather than improved, out well-being. While we may still try to convince ourselves that more material wealth and consumption improves our well-being, all of the evidence suggests the opposite. In his enlightening study, *The High Price of Materialism*, psychologist Tim Kasser points out (based on extensive peer-reviewed research), that as our wealth increased, our well-being actually decreased! The results of the research suggested "a rather startling conclusion: the American dream has a dark side, and the pursuit of wealth and possessions might actually be undermining our well-being."(p. 9) Kasser concludes that "When materialistic values dominate our society, we move farther and farther from what makes us civilized."(p. 91) Despite these conclusions, confirmed by extensive research, we continue to convince ourselves that our only path to happiness is one of developing the capacity to acquire as much material wealth as possible.

Enter Michael Rosmann and this lovely book. What Michael, a psychologist and farmer, reveals to us through a treasure of personal stories is that true fulfillment comes not through a "gospel of consumption," but rather through a gospel of community. The rich experiences that bring us true fulfillment are those that we encounter in simple communal and family relationships, relationships that develop during simple acts of fishing and hunting and community-based farming. While Michael may have been prepared to come to some of these conclusions through his academic training in psychology, he reminds us that he is "a better psychologist" because of his "encounters with nature," and he encountered these relationships through hunting, fishing, and farming. As his stories reveal, these encounters are what rewarded him with the true fulfillment of "excellent joy."

Accordingly, the stories in *Excellent Joy* are much more than simple stories of one man relaying to us some of his life's experiences—they are stories which reveal the power of the gospel of community—and not only the community of humans, but the community of nature as well. As Michael reminds us, his experience as an observant farmer taught him that tame and wild need not be separated on a working farm, because "rich farmland produces bounty for wild and domestic alike." As Aldo Leopold reminded us, the "separation of tame and wild" exists only "in the imperfections of the human mind," a discovery that for Michael, produces yet another "excellent joy."

—Frederick Kirschenmann

Frederick Kirschenmann is the author of *Cultivating an Ecological Conscience: Essays from a Farmer Philosopher*. He is also Distinguished Fellow at the Leopold Center for Sustainable Agriculture at Iowa State University and the President of his family's 3,500-acre certified organic farm in south central North Dakota. He helped to found Farm Verified Organic, Inc., a private certification agency, and the Northern Plains Sustainable Agriculture Society and has served on the USDA's National Organic Standards Board, the North Central Region's Sustainable Agriculture Research and Education (SARE) administrative council and the Henry A. Wallace Institute for Alternative Agriculture board of directors. He is the Board President of the Stone Barns Center for Food and Agriculture. Dr. Kirschenmann won the National Resource Defense Council's Growing Green Thought Leader award in 2010.

Acknowledgements

This little book has been a long time coming. I first began thinking about recording formative life experiences in print about twenty years ago. Indeed, I wrote *The Meaning of Christmas* in 1991. It was one of my first ventures into creative writing since my graduate school days. I became determined to tap my feelings for written inspirations, something I set aside when professors drilled the scientific method into my head.

I owe a great deal of appreciation to my family, my staff, and to several people who constructively criticized my writings. First and foremost, I wish to thank my beloved wife of 39 years, Marilyn. She often encouraged me to put my thoughts into print. She put up with me arising from our bed in the middle of night to jot down ideas. She was, and still is, a positive and grounded sounding board.

In a sense, this book should be dedicated to Marilyn's father, Walt. His influence on me is interspersed throughout the book. He taught me how to fly fish. Walt set a wonderful example about how to find joy from simple possessions and activities. We became good friends, fishing buddies, and colleagues in life.

I am grateful to my mother-in-law, Michi, for allowing Walt the freedom to express his pursuits of the outdoors and for instilling in Marilyn many of the desirable characteristics that have made my life joyful. Although the ends of their earthly lives have come to pass, the spirits of Walt and Michi live on in us.

I tried to pass along to my children, Shelby and Jon-Michael, zest for working and playing hard and the parent-child bonding that I acquired with Walt. The apple didn't fall very far from the tree—it fell right into

Jon's lap. A family joke is that he was born holding a fishing rod. He could catch his share of bluegills at four years of age, even with a bare hook. He loves hunting and fishing, the outdoors, and going with the flow. We are good friends as well as hunting and fishing buddies and are fully honest and respectful of each other. Amanda, Jon's bride, fits right in with this outfit. Amanda loves Nugget, Jon's hunting dog that you will read about frequently in this book, nearly as much as Jon does. She likes big gardens, her own two dogs and three cats, and outdoor ventures.

With Shelby the apple took root in different forms. Shelby began taking care of people as a child and now as a physician. She rode with me in the moving van when we made our trek from the academic community of Charlottesville, Virginia, to our farm in western Iowa in May 1979. It was a grueling 1200 mile drive through rain nearly the entire distance. Shelby, then four years of age, helped me fight off fatigue by singing sustaining songs—she postponed her own naps to help me stay alert. Over the next couple years, when she didn't have to attend school, Shelby rode with me in the tractor and combine cab as we farmed. She was in charge of the food and entertainment. She took it upon herself to get our camper ready whenever our family had an opportunity to take a break for camping, fishing, and exploring the outdoors. Shelby and her husband, Shale, love the outdoors and they now live in the mountains near Salt Lake City. I am grateful that she brought Shale into her life and ours. He regularly admits his favorite recreational activity is coming to Iowa to hunt pheasants and turkeys.

My staff at AgriWellness, Inc., a little nonprofit corporation devoted to helping farm people with behavioral health issues, has been absolutely wonderful. Carol, my administrative assistant, has typed every one of these stories, some several times. She goes about her tasks deliberately, with dedication and drive. Sometimes I hear her laugh when something I have written strikes her as funny. I thank Shari for being a patient critic and helpful reflector of what might be appealing to readers. She is a good proof reader and a trusted colleague who tells me what I need to hear rather than what I might like to hear. Linda, who died three years ago of cancer, also typed many of the earlier versions of these stories. The story, *Tachycardia,* took place on the farm where she and Bob raised their family. We all still miss her at AgriWellness. Before Carol joined us, Deb typed stories and I remember her contributions.

The well known author, Dan O'Brien, commented on a number of these stories and his feedback and tutelage helped me. My writings probably never will be as well known as Dan's but his examples are good ones for me.

I asked several people to read the stories and to tell me their impressions. Thanks go to my friend, Al, and especially to Steve Semken, the publisher of this book. I appreciate Amanda's father, also named Steve, for his photo of the little black fly.

I am grateful to all the people who inspired the stories in this book. Occasionally to protect the innocent or the guilty, I changed a name here or there. The events themselves happened pretty much the way they are described in the book. That's what has made them such excellent joy.

In closing, I want to thank my deceased brother, Larry, for the title of this book, and more. Eight years younger than me, Larry experienced Down syndrome. He had a limited vocabulary, but whenever people or events were especially pleasing to him, he uttered in his raspy voice, "Excellent joy!" Although Larry had disabilities, he "got it together better" in life than me and most people. He was a good person, strongly spiritually motivated, and he cared about others. Larry was always happy and content with small things in life. *Excellent Joy* is a fitting title for this collection of stories about fishing, farming, hunting, and psychology. I hope readers get as much fun out of these episodes as I did writing the stories.

—Michael R. Rosmann

You're the Only Guy I Brought Home That Daddy Invited to Go Fishing

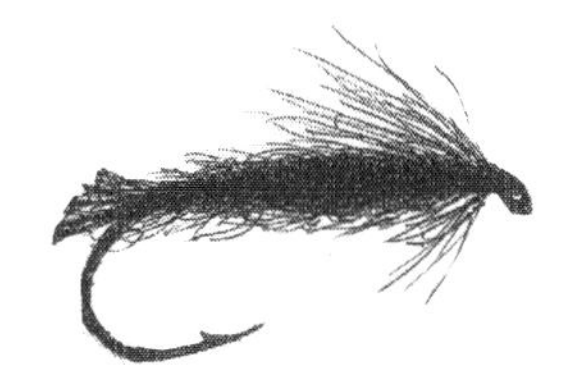

"You sit in the middle and let Walt have the window," Digger commanded as Walt and I stowed our fishing gear in the back of the pickup and secured the rear door of the aluminum topper on Digger's truck.

"Mike can sit by the window. He's got longer legs than me and needs the room," Walt suggested.

"No, he's a young rascal and more limber than us old farts," Digger rejoined. I slid into the middle of the bench seat, one leg on each side of the transmission case and held my breath every time Digger shoved the gear lever straddled between my legs. I hoped Digger's hand wouldn't slip off the transmission lever when he shifted gears.

It was June 1971, about four months after Marilyn and I met. This was our second weekend trip to her Burley, Idaho, home from Salt Lake City, where we were graduate students at the University of Utah. Walt and Dick "Digger" Philson invited me to go fishing with them on a relatively undiscovered river from which water was extracted to irrigate the potato and sugar beet fields in southern Idaho. Digger was a county commissioner and ran the local mortuary and he often fished with my future father-in-law, Walt.

I brought with me the only fishing equipment I owned: a casting rod and reel, a wooden handled net, a few hooks, and a jeweler's pliers my mother gave me several years earlier. I didn't own a pair of waders. Walt and Digger liked to fish a stretch of water below the irrigation canal inlet where the water had a reduced flow that I came to learn was called, *"the cottonwood hole."* Years earlier Walt had introduced Digger to fly-fishing, a pursuit I had only read about.

Since all my fishing experience was with bait-casting, the evening before, Walt and I caught a couple dozen night crawlers on his lawn, using a flashlight with red cellophane over the lens after Walt had turned on the sprinkler for an hour. As Walt and Digger waded into the river in their chest-high waders, Digger advised, "You can walk along the shore below us and throw your garden hackles into the holes I show you."

From my downstream vantage point I envied the graceful arc of Walt's fly line as he delicately looped his rod over his head and with a vigorous forward motion of his right arm thrust a hand-tied fly, 60 feet ahead of him, into the tail end of a deep pool. Years of floods had scooped out a depression a couple hundred feet long. After a few-second pause, Walt began to retrieve line. Suddenly the line became taut. Walt quickly raised his rod. The front half of his rod bent downward and whipped back and forth. A hundred feet upstream an eighteen-inch rainbow trout leaped three feet out of the water. Patiently, Walt let the fish course back and forth across the pool but always he kept his line tight. Gradually Walt retrieved line through his right hand and stored it in coils between his left thumb and forefinger. Periodically the fish unraveled some of the coils but Walt carefully let the fish make a run and then slowly regained the line until the fish was a few feet in front of him. Switching the line and rod into his left hand, with his free hand, Walt grabbed a cork handled net attached to his fishing vest and swept it under the tired fish. I thought, "*I would like to do that.*"

"What did you catch him on?" I asked loud enough for both Walt and Digger to hear.

"The Big Shrimp Special," Walt replied.

"Yeah that's the hot one here," Digger solemnly pronounced. Walt had given Digger several of his specials.

By lunchtime Walt and Digger had captured many rainbow trout but the two men kept only four rainbows apiece, all 16 inches or longer and I had nothing to show for my efforts. After wolfing down a roast beef sandwich I cast a hefty worm into the riffle next to where my companions were still eating. As I cranked the bait toward me I felt something pull hard on my line. The tip of my rod thrust downward. What appeared to be a heavy fish swam strongly downstream in the riffles ahead of me. I scrambled along the riverbank to accompany the fish. Walt yelled, "You got a good one!"

Digger joined in, "You better catch it, it looks pretty good."

My rod bent considerably. I reeled furiously but gained little line. It was a good thing I had used 15-pound test line, because the fish managed to pull several rolls off the spool even though I set the drag tightly.

"*Great,*" I thought, "*This has to be the biggest trout of the day.*" As the fish's stamina flagged, I was able to draw the catch toward me. I was sure I had hooked at least a 24-incher, given the encouraging words of Walt and Digger. It came as quite a disappointment when I dragged a 15-inch rainbow trout onto the shore at my feet. To make matters worse, I had hooked the biggest trout in my 24 years of life in its tail!

As the fish flip-flopped on the cobblestone shoreline, Digger loudly summed up the situation, "Foul-hooked fish don't count."

Walt countered, "You caught it and you should keep it." I reached down to remove the hook from the tail of the now still trout. As the hook pulled free, the rainbow twisted violently and quickly flipped into the shallow water nearby. I lunged desperately to seize the fish but the slippery body escaped my hands and quickly darted into the dark deep middle of the stream.

I wanted to cuss but I didn't want to signal my fishing buddies that I could be a bad sport. "Doggone, I'm going to have to learn to fish like you guys."

At day's end, and after reciting our reports to the women waiting at home, Walt invited me into his basement. He displayed his fly tying bench, littered with numerous cigar boxes stuffed with feathers, fur and hair of all sorts, a multitude of little trays in a battery of boxes that contained hooks, paint, chunks of wax, and other things that I had never seen before. There was a vise clamped to the front of the workbench with a bright light bulb suspended directly above it. Spools of thread, several scissors, awls, clamps, and tweezers were strewn on top of the bench. Pheasant tail feathers, duck wings, squirrel tails, and pieces of chenille stuck out of cabinet drawers on the desk. Walt showed me fiberglass and bamboo rods that he was repairing for other people. Half the basement was filled with waders, nets, bunches of rod holders tied together with twine, boots, canvas covered float tubes, truck tire inner tubes, Styrofoam coolers, and wicker creels in various degrees of fading.

"Sit down," Walt said as he pointed to the stool in front of the fly tying vice. I was enthralled as he guided me through the steps of tying my first fly, a Big Shrimp Special.

"Walt is showing me how to tie his favorite fly," I thought to myself, *"and he hardly knows me."*

I had already learned three lessons on this day: 1) I wanted to become a fly fisherman, 2) foul-hooked fish pull deceptively harder than fish properly hooked in their mouths, and 3) unhook fish only after they are in the net or on a stringer if you want to keep them. Now I wondered, *"What lay ahead? What is this man like?"* I sensed that I could learn more from Walt about what is important in life than I was acquiring from my graduate school psychology professors. *"It would be fun to get to know Walt,"* I reasoned, *"if Marilyn and I marry."*

I felt I knew Marilyn's mother, Michi, better than Walt because Marilyn often described how she expected Marilyn to share intimate details about the men she dated, whether they prayed, smoked, bathed daily, worked hard, and were honest. Marilyn commented that she wanted to marry a farm boy with a PhD. Michi apprised her that if she married me she would have trouble having our babies, "Look at his shoulders!" Michi was impressed that I washed the dishes after meals and liked every kind of food.

I had not learned much about Walt except that he wasn't a Seventh-Day Adventist like Michi and that he fished at least twice a week and often more. It was clear that Michi was in charge of their relationship. Walt was quiet. Fellow fishermen called him nearly every day to arrange outings but Walt often fished alone. Maybe he fished to experience solitude, but it was probably more than that. I wondered, "*What is it that makes everyone want to fish with him? It has to be more than to receive his generously-given flies.*"

"Where did you learn to fly fish?" I ventured, while putting the final dabs of head cement on my first fly. This started the first of many pleasant discussions over the next 35 years, most while riding to and from fishing haunts and eating shore lunches together.

Walt was born near Selma, California, 33 years before I was born. The child of immigrants from Japan, Walt grew up on a farm where he and his family raised golden seedless raisins, avocados, and citrus. He graduated from Selma High School in 1930 but deferred going to college so his younger siblings could attend college. Walt told me he learned to fish first for bass in nearby farm ponds. He still had bass plugs he constructed from balsam wood. He read about fishing with flies from books by Zane Grey and built his own bamboo fly rod from a kit he purchased out of a mail order catalog. Walt experimented with fly fishing in the Sierra Nevada Mountains east of Selma. Sometime in the 1930s he penned in his journal:

> There is no creature as strong in proportion to its size and weight as a trout. Its home is in the icy torrents that are fed by the snows of the highest peaks and canyons. It lives literally in the innermost heart and life of the hills. It seeks its food at the foot of the hills where the water boils in fierce fury, where the current swirls and leaps among the boulders and where the stream rushes with might down the rocky channels. With muscles fine as tempered steel, the trout pushes its way against the constant current, conquering even 20-ft. cataracts. Its strength is silent. It has no voice other than its existence.

After World War II broke out in 1942, Executive Order 9066 required people of Japanese ancestry on the West Coast to relocate to inland camps. Walt first went to Manzanar, California, and then to Gila River, Arizona. Two weeks after Walt entered the relocation camp, his barn, his car, a boat, homemade skis, and all his possessions except the clothes and fishing gear he had with him mysteriously burned in a huge conflagration of the barn where he had stored them. He wasn't bitter. Fortunately, Walt kept his favorite fly fishing reel with him that he used all the rest of his life, a Pfleuger Medalist he purchased in 1933. Most of its black enamel had worn off but it worked oh so smoothly.

Relocation camp days forever changed Walt's destiny. At Gila River Camp in Arizona, Walt heard about a good looking single nurse working in the infirmary and arranged to have a convenient cut on his hand examined by his future bride. The 59-year marriage of Walt and Michi resulted in Marilyn and her two younger brothers. After transfer to Minidoka, Idaho, Michi was released from camp to take a job as a nurse at the new Burley hospital.

At first Walt did the next best thing to farming—gardening. Then he repaired fountain pens and electrical appliances for many years. Eventually the local potato processing plant hired him as an electrician and mechanic until his retirement. Idaho provided Walt the opportunity to shape his fly fishing and fly tying skills to become an expert, but he never let on about that. Even when he hooked a fish or lost a big one he said nothing, but when we compared catches at the end of the day, Walt nearly always had the biggest and most fish. Like Digger and the doctors who worked with Michi and fished with Walt, I learned I could tell Walt of my accomplishments, flubbed cases, and worries and never had to be concerned he would repeat what I told him, not even to Michi. He cast no judgments except to say, "Good," or "Try harder."

Two weeks later Marilyn and I were back at her parent's home. I was now the proud owner of a fiberglass fly rod, a Medalist reel, and chest-high waders. Walt and I had planned to go fishing on both Saturday and Sunday, using float tubes. When Marilyn and I pulled into the driveway, Walt was in the carport next to the house pumping air in two truck tire inner tubes with attached canvas seats. Two sets of flippers lay on the concrete driveway, "Will these fit you?" he asked, pointing to a new set.

Delighted, I thought to myself, *"Marilyn must have told her mother in one of their many weekly telephone calls that I was quite taken with her father and had purchased a fly rod, reel and waders. Michi had passed the word along to Walt."*

When I commented to Marilyn how surprised and delighted I was, she replied, "You're the only guy I brought home that Daddy invited to go fishing."

Idaho Crappie Fishing In Winter

Before I met Walt I thought that crappie fishing in winter was—well—crappy. The only way I thought one could fish for crappie during winter was through the ice and usually the catching part was slow. Walt and his friend, Bert, took me with them to fly fish for crappie in southern Idaho in the middle of January the first year after Marilyn and I began to see each other. I figured they took me with them partly because it might be good to get to know the future son-in-law. Kind of like the movie *Son-in-Law* in which the daughter's family and acquaintances decided it was their duty to educate the newcomer so that he might fit in a little better.

Walt was one of the better fly fishermen in Idaho and one of his favorite fishing buddies was Bert, the barber. Bert did not cut Walt's hair though because Walt always cut his own hair using two mirrors. Bert had a boat and I could ride with him in the boat while Walt used his float tube to fish.

For years, Walt and Bert often fished at a little reservoir fed by a somewhat sulphurous but hot spring. Typically, the channel in the reservoir was free of ice for about a half mile downstream from the spring, while the rest of the impounded water in the reservoir was otherwise frozen thickly solid on the surface. However, if you could get your boat or float tube into the open water, you could work your way up and down the channel, picking off crappies that hung out in the submerged willows on either side of the channel. All you had to do, Bert said, was drop a fly into the water next to the edge of the ice, let it sink toward the base of the submerged willows, and wait for a hungry black crappie to snatch it as it descended.

It was a cold, clear, late Sunday morning by the time we reached the reservoir because Bert's wife always made him go to church before she would allow him to go fishing. The county road maintainer had dutifully pushed the snow off the gravel road leading to the reservoir and had cleared the parking lot and boat ramp. Bert backed his pickup truck and boat trailer onto the sloping concrete ramp where the channel approached the shore. Walt and I unhooked the boat. As Walt prepared his float tube and fishing equipment, Bert and I loaded our gear into his battered, 12-foot aluminum boat. Bert showed me how to tie two gunny sacks onto the side of the boat so they would hang into the water to keep the fish fresh throughout what we both expected was going to be a pleasant and bountiful winter harvest.

Bert and Walt were telling me that they regularly caught 30 to 40 crappies apiece on a good afternoon when a big SUV, pulling a nice shiny red and white bass boat, pulled into the parking lot. Bert threw two six-packs of Rocky Mountain kool-aid into his banged up aluminum craft while the two recent arrivals sauntered over to talk. They kept saying, "eh" after every sentence. It seems they had heard about fishing for crappie in Idaho in the dead of winter and decided that they would give it a try while heading from British Columbia down to Mexico for some mid-winter bass fishing.

While casually eyeing Bert's dented boat, Walt's float rig, and all the fly rods, one of the Canadians ventured to ask, "What do they hit on, eh?"

Not wanting to appear unkind to the apparent fishing neophytes, Bert suggested, "You've can catch 'em on a fly but you've got to attract 'em first."

That seemed to spark some deep thought in one of the Canadians, for he stammered that he had heard somewhere that if you beat the water with your oars, crappie would come to see what was causing the commotion and then you could catch them. With a twinkle in his eye, Bert volunteered that this was probably correct.

By now Walt was already paddling his float tube up the channel. Bert was in his boat. I shoved off and jumped into the boat with him. We headed in the opposite direction from Walt so as to try to cover the channel while the Canadians figured out how to catch crappie.

For the next two hours Bert and I drifted down the channel and Bert picked off one crappie after another. Each time his rod bent in the shape of an inverted U, Bert laughed and gulped two or three swigs of beer. He tossed each eight to ten-inch crappie to me to put into his gunny sack. Despite the fact that Bert spent a good deal of time emptying and refilling his radiator, he caught at least three crappie for every one that I landed.

When we reached the lower end of the channel, where ice was forming on the surface, Bert turned the boat around and worked back toward the boat ramp. As we rounded the last bend in the channel and could get a clear view of the open water by the boat ramp, we heard a noisy commotion and witnessed something neither of us had seen before and which I have not seen since: two grown men were leaning over the sides of the red and white bass boat, furiously flailing the water with their oars. Bert started to snicker, then broke into a deep belly laugh. His laughter was contagious. Finally Bert belched and guffawed so loudly that I don't know who I was laughing at more—Bert or the busy Canadians churning the water surface into a bubbly froth. Bert gave the outboard motor a little more gas to get around the comical sight and headed upstream toward Walt.

As Bert and I drew within sight of Walt, Bert started to snort and laugh hilariously again. Walt chuckled almost as loudly as Bert filled in the details of the scene he and I had witnessed.

What an entertaining afternoon, I thought to myself. "*Here I am, fishing with two fine, Idaho fly fishermen and while I probably caught only nine or ten crappie, Bert and my future father-in-law were initiating me into their ranks as fellow fly fisherman and someone who is privy to the inside secrets of good fishing.*" The merriment caused by the misguided Canadians added substantially to the luster of the event.

Bert suggested that as nearly all the liquid nourishment in the boat was gone, and we had caught enough fish, perhaps we should head home. When Bert asked Walt how many crappie he caught, Walt held up two metal chain stringers with two fish on most of the stringer snaps. Then Bert suggested to me, "Why don't you show your future daddy-in-law how many we caught."

I hoisted my gunny sack into the boat and dumped out ten or so black and white speckled fish, making sure that Walt saw each one fall from the sack onto the boat floor. Then I passed Bert his bag so he could raise it out of the water to reveal his catch. Beaming, Bert raised his bag but his arm rose entirely too quickly and easily. As he pulled the nearly flat bag into the boat, two salt-and-pepper colored fish flopped to the boat floor. Bert had forgotten that he had not repaired the hole in the bottom of his old gunny sack creel. He sputtered and spewed such a rich variety of phrases that had my mother ever heard me utter them, she would have made me chew on a Lifebuoy soap bar. I lamely consoled Bert, "That's too bad."

Becoming subdued, Bert mumbled "I didn't want to show up anybody."

Now I slowly headed the boat toward the ramp and Walt paddled gamely to keep up. Bert had put me in charge of the motorized transportation because he was in no shape to drive the boat after losing his fish, but the last can of beer was helping him regain his composure.

As our boat approached the concrete incline, the two Canadians were going about the finishing touches of securing their fancy red and white bass boat to its trailer. Feeling a little cheered at seeing the Canadians preparing to leave and working hard to suppress a chuckle, Bert ventured, "How did you boys do?"

I turned my head away so our northern neighbors couldn't see or hear the raucous roar that I was trying to stifle. "Not too bad, for our first time," one of the Canadians answered.

"Let's see your catch," Bert managed to gasp. From under the canvas cover of the boat, the gentlemen arduously lifted two wet bulging gunny sacks, each as big as a twenty pound bag of potatoes. The obvious movement of the gunny sacks indicated that whatever was inside was alive and there were a lot of them.

The Little Black Fly

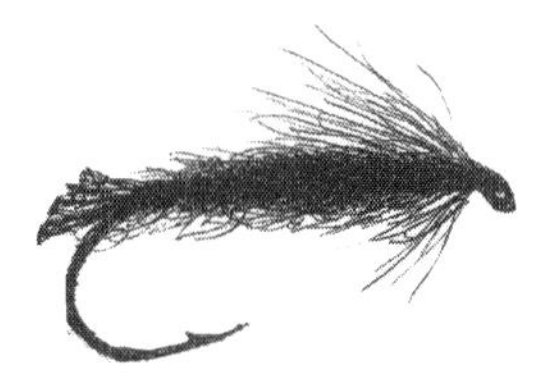

My wife and I moved to Charlottesville, Virginia, so that I could take a job as Assistant Professor of Psychology at the University of Virginia in August 1974. I had finished my PhD in clinical psychology at the University of Utah. The University of Virginia was considered one of the most hallowed spots in the country to begin an academic career. I was a bit scared and uncertain but my parents, Marilyn, and her folks were immensely pleased. The boy had made good! Not bad for an Iowa farm boy whose parents had not gone to college.

By then I was starting to learn a thing or two about fly fishing. Earlier, while a graduate student at the University of Utah, I figured out how to crouch low and float small dry flies or single artificial salmon eggs with the current, under willow-shaded small streams tumbling down the Wasatch Mountains east of Salt Lake City. I would set the hook just right when hungry cutthroat and brook trout slurped the bait. My roommate, Ken, used bottled salmon eggs, worms, or begged flies off me while I used hand-tied nymphs to explore for stream-born brown, stocked rainbow, and native cutthroat trout in the Weber, Duchesne, and Provo river drainage out of the Uinta Mountains.

I camped a lot, sometimes with Ken or Marilyn, both, or occasionally alone. Ken had introduced me to Marilyn, something for which I will always be grateful. Ken and I traveled to the Madison and Jefferson rivers in western Montana a couple times and I fished just outside of Ennis by myself several times. When Marilyn and I camped our way through Idaho, Montana, and British Columbia on our early autumn honeymoon, Marilyn chided me to be careful as I struggled to stand upright in fast flowing wild waters and drove off black bears with rocks. I learned of superb trout lakes on the Duck Valley Indian reservation in Nevada with Walt and Digger. My father-in-law and I, and sometimes one or two of his fishing buddies, sought out crappie and bluegill, but mostly trout on many southern and central Idaho lakes and streams. Walt's tutelage was a significant influence on me. We developed comfortable, mutual respect, and boundaries in our relationship during our many trips to Idaho reservoirs and rivers. Some of his tutoring helped me later to set the tone in my relationships with my two children.

My father rarely hunted or fished. My family never camped and seldom took vacations, for that matter. My parents were third generation German immigrants and products of the Great Depression. As farmers, the land meant nearly everything to them. I learned how to work efficiently, persistently, and not to quit until the job was finished. Those traits were so engrained in me that I could outlast what I perceived as hazing, and sometimes unreasonable demands, from graduate school faculty. I stuck it out even when my professors made me rewrite my master's thesis 19 times.

I had set out to become a competent clinician, a good teacher, and scientist. I had not learned the necessity for recreation during my childhood. By my mid-twenties I was gaining understanding about how to work and play hard too. When I was old enough and confident enough to engender an adult relationship as an equal with my father, he died of a massive arterial blockage in his heart. I would have liked to learn more from him and enable him to enjoy life more when Marilyn and I moved to Iowa to farm after a five-year stint in Virginia. Dad and I spent only a year together in Iowa and were just beginning to experience the fruits of working, talking, and playing together when the tables turned. Instead of Dad and Mom sharing their accumulated knowledge with us, we had to bury Dad and take care of Mom. It was especially sad that Dad did not catch a fish during his entire life. Three of us took a trip together once to Kenora, Ontario. The rest of us caught all the northern pike and walleyes we could legally keep, and Dad cast the same lures we did, nearly as much as the rest of us, but he never got to experience the tug of a piscine prey at the end of his line.

When Marilyn, very pregnant, and I moved to Virginia, I began to think differently about fly fishing, hunting, and about life and responsibilities in general. We bought a three-acre lot with an acre-sized pond brimming with bluegills and bass a few miles west of Charlottesville on the same farm where Meriwether Lewis had lived. We decided to build our first house by ourselves as much as possible. Our first child, Shelby, was born shortly before Christmas that first year and now I had a family to feed. Academic work was rewarding, but the salary less so, making money tight. I planted a huge garden and groomed wild Albemarle pippin and black cherry trees already existing on our property. Marilyn and I canned or froze most of the fruits and vegetables we raised, sometimes staying up to the wee hours of morning finishing a batch of sweet corn or cherries. Marilyn sewed all the curtains for our new home, most of her clothes, many that Shelby and I wore and later Jon's as well. Jon-Michael, our son, came along three years after Shelby. I think he was born with a fishing pole in his hand. We were having a lot of fun as well as working hard.

I fished and hunted, now, not just for the inherent joy of these pursuits, but also to put food on the table. We ate much venison, geese, and fish of all types.

Drawing on the essentials of fly tying that I learned from my father-in-law, I experimented with many colors of fur, feathers, and fabrics on the brook trout streams of Virginia, nearby West Virginia, and the many bream-filled streams and lakes around Charlottesville—Beaver Creek Reservoir was a particular educational opportunity.

Beaver Creek Reservoir, only three miles west of the house that we constructed, was located in the foothills of the Shenandoah Mountains. The very pretty willow and pine-ringed, two-mile long reservoir was my training arena. I also tried out my fly-tying concoctions on the one-acre pond on our homestead but I didn't want to remove too many fish from this small fish nursery. So my family and I often ventured to lakes and streams within easy driving distance, and especially to Beaver Creek.

The first year Marilyn and I lived in Virginia I built a canoe of wood and fiberglass, with the assistance of a UVA colleague and following directions from a *Popular Mechanics* article. It's a beautiful wood-grained craft that still is operational after 35 years of plying streams and lakes throughout the country and running a great many rapids. Marilyn proved herself to be a pretty good motor on the canoe, which allowed me to fish full-time unless we were trying to get somewhere in a hurry. Marilyn wasn't interested in fishing herself, she just liked being outside. The kids went with us. As we trolled Beaver Creek Reservoir, I noticed that minnows seemed to congregate in weeds and fallen tree limbs near the shoreline. I cast various flies next to the debris, such as streamers that I had acquired from Walt or tied myself, using patterns from fly fishing books and magazines. I purchased a few minnow imitations from mail-order catalogs. Sometimes they worked and sometimes they didn't.

When I spotted slender yellow-green caterpillars, crawling on alder branches leaning over the reservoir, I created a pattern that worked fairly well during the summer as long as the errant inchworms sometimes fell into the water. But I had no minnow patterns that produced regularly. I tied flies with gray or black chenille bodies and black feather hackles, but when I tried them they were too bulky and seldom resulted in strikes. I put together flies made of black or gray goose quills and black rabbit fur with some positive results. I launched an investigation of available options. I was learning how to experiment with a variety of materials and hook sizes to approximate minnows.

After multiple fly fishing experiments, I eventually determined that something entirely new was needed to resemble the minnows that predator fish seemed to relish. The fly had to sink slowly, dart deceptively when retrieved, and display fin and tail action. I tried various materials for the body and fins. Black seemed to be the best choice of color. Black wool yarn or thread seemed to create the best shaped body. Black mink fir seemed to have the optimal action for the fins and tail. A small amount of weight on a size eight hook increased the hooking rate.

Eventually my invention took on proportions that were somewhat symbolic of my life: many experiments, failures, and trying again until something worked! Now almost everyone who uses the fly affectionately refers to it as, "the little black fly." At least I think that is what they are saying!

I found that tossing almost any fly into colonies of bluegill nests would result in periodic hook-ups, but the little black fly always seemed to trigger harder and more consistent strikes than alternatively colored flies of the same size and shape. Whether along rocky shores, patches of weeds, moss, or in the deeper and cleaner interiors of ponds and lakes, the little black fly seemed to have better payoffs. It enticed grumpy, big old largemouth bass on hot August afternoons. It instigated feeding frenzies among schools of crappies until the fly was worn down to just a hunk of thread and a few black hairs clinging to the hook. The fly landed a 24-inch walleye on Merritt reservoir in north central Nebraska, on a cold, windy, cloudy May day when nothing else worked and nobody else was catching anything.

In early April 2009, Jon, Scott, and I made our annual trek to northern Arkansas to fish the Norfolk River below the dam. While Scott and Jon floated the river in pontoons, I parked my Jeep and trailer five miles downstream to pick up the guys at day's end and to fish while I waited. Jon and I kept in touch via our cell phones. I landed a couple decent fish on a sow bug and a red-butted green nymph, but the strikes were coming a little slowly. Just then Jon called, "Try the little black fly. I caught three on it already."

I tied on a size eight little black fly and hooked a 16-inch rainbow on my first cast and a 17-inch cutthroat on my second cast. It was the only fly I used for the rest of the trip. Who would have thought! I caught my limit of five fish every day during the rest of the trip and released three-four times as many more.

Everywhere others have tried the little black fly, its magic has worked—on largemouth bass, white bass, crappies, sunfish, bluegills, perch, and walleyes. A couple years ago Jon caught two seven-and-a-half pound channel catfish on the black fly. The fly was successful in North Carolina lakes, Midwestern streams, and farm ponds and even in a dinky, mossy, spring-fed cattle pond in western Idaho. Of course some people want to know how to tie the little black fly. Here goes.

Start with a Mustad 3906B or similar hook. What sized hook? It's good to have a selection of sizes ranging from No. 12 to 4, depending on the size fish you are trying to catch and their habits. Generally, No. 8 and 10 are the best all around because they imitate the size and movements of young fry. The larger sizes help avoid catching smaller fish, which can be quite a nuisance when fish populations are high. Small fish can't get the larger hooks in their mouths. There have been times though when a No. 12 has consistently caught more big fish than a No. 6, probably because the minnows that big fish were feeding on were still fairly tiny.

Place the hook in the vise and put 8 or 10 wraps of thin lead wire on the shank of the hook. The lead wire and the hook itself give the finished fly enough weight to offset the natural buoyancy of mink hairs. Then, starting from the eyelet end and working backwards, tie down the lead and make several revolutions around the back of the shank where the hook starts to curve. Use black waxed thread or mono-cord. It's helpful to put a little dab of cement on the rear wraps to keep them from sliding on the hook shank.

The next step is to select and tie 15-20 black mink hairs onto the rear of the shank to form a tail. The hairs should be long enough so that the tail extends 1/4 inch beyond the point where they were tied to the hook for No. 12 flies and about 5/8 inch or longer for No. 4 flies. Tie the tail hairs so they don't fan out excessively and can form the tail end of a fry as the fly is pulled through the water. If natural black or dyed black mink hairs that are long enough are not available, black deer hairs cut and tied to correct lengths can be substituted. Use several thin strands of dyed black wool to form the body of the fly. As the wool strands are wrapped, they should form the shape of a tiny fish body, tapering toward the front of the hook to form the head. While black wool thread

> works best, other materials which have a slight sheen, such as black poly yarn or Z-lon, can be used satisfactorily.
>
> Now the fly is ready for the development of pectoral fins. Again, mink hairs, or a substitute, should be used and they should be about the same length as the hairs that were used for the fly tail. Gather about 25-30 hairs and tie them to the top half of the fly where the front of the body ends. This shape allows the hairs on the top and sides of the body to flex and to appear like pectoral fins as the artificial minnow "swims" through the water. Make a few wraps of thread to hold down these hairs and to form the mouth of the artificial minnow. After cementing the head, the fly is finished, except for trying it out.

I felt especially good that Walt had a lot of success with the little black fly. He used the fly effectively before he died, which helped me feel I was repaying my father-in-law for the ample knowledge that he imparted to me. Two years before he died, Walt contacted an Idaho mink farmer with whom he was acquainted to obtain a big bag of black mink tails and leftovers from assembling fur garments that will last me the rest of my life, even if I tie a couple hundred little black flies every year. Hardly able to get around, the last time Walt went fly fishing he sat in a lawn chair next to a farm pond casting the little black fly and landed the biggest bass of the day.

Rhododendron Culture and Fly Fishing

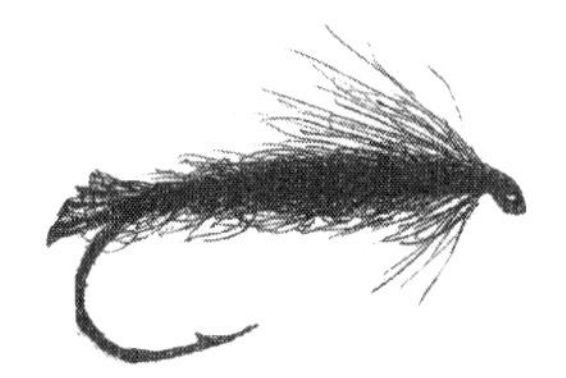

It was February and trout fishing season was closed in Virginia. I had suffered the past few months. With fishing and hunting seasons closed and the soil too frozen to work in the garden or on the lawn of our new home, I poured myself into my job as an Assistant Professor of Psychology at the University of Virginia. I worked 70 - 80 hours a week at my university office. At home I invented children's stories and played carols on my guitar to entertain Shelby and Jon. I helped with the cooking, laundry, and household chores as much as I could. Yet, I still couldn't sleep easily on some nights. Like Walt, when I couldn't indulge the real thing, I did the next best—either I tied flies or read about fly fishing.

Over the past couple winters I read every book on fly fishing—about a dozen—in the University of Virginia Library. I decided it was time to point out the deficiency of fly fishing literature to library officials, so I penned a letter to the Library Acquisitions Committee:

The University of Virginia
Library Acquisitions Committee
Charlottesville, Virginia

Dear Library Acquisitions Committee,

Every winter I suffer a terrible affliction. I am addicted to fly fishing. There, I said it! When fishing season is closed during the winter, the only way I can assuage my addiction is to read about it. I have read every book about fly fishing available in the University of Virginia Library.

Would you kindly please look into purchasing additional books on fly fishing for the library? Here are some suggestions: [Here I listed 20 titles, authors, and publishers of books about fly fishing and I put asterisks by those that I thought deserved preferential consideration.]

If you could help this poor miserable addict, I would be forever grateful.

Very sincerely yours,
Michael R. Rosmann, PhD
Assistant Professor

About two months later, while taking a walk on our country road, I happened to run into our next door neighbor, John Grubbs, the associate librarian at the University of Virginia Library. We had often exchanged, "hello," compared our kids and hobbies and discussed the finer points of John's great love when he wasn't sorting books: rhododendron culture. John had tenderly nursed rare rhododendron specimens all over his three-acre wooded homestead. He had even carefully removed 100 year-old oak trees to create avenues of sunlight to penetrate down to the multitude of rhododendron shrubs he had obsessively spaced in manicured plots. A pink one here, a violet one there, in that corner a whitish flowered shrub all the way from Ireland. Usually I was unenamored with these discussions, but I listened patiently because John was as enthused about rhododendron culture as I was about fly fishing. As we approached each other on this day, I observed John's face turning red and his cap seemed to lift off his head as his hair was beginning to rise to attention.

John spoke first, "Did you get my letter?"

"No, I haven't received a letter from you," I responded.

"Well," John sputtered, "I sent it to you yesterday. I have been trying for more than ten years to get the Library Acquisitions Committee to approve a single book on rhododendron culture, to no avail. Then you wrote only one letter requesting 20 books and they approved every selection."

"Great," I said. But, seeing John's face and neck grow even redder and his cap elevate even farther, I thought, I should temper my enthusiasm, so I asked, "How did that happen?"

John responded, "It seems that a number of years ago somebody left $2000 to the University of Virginia Library with the condition that the money must be spent on books about fly fishing."

"At least somebody has the right priorities," I silently thought to myself. I could see that John was enormously exasperated and reckoned it might be diplomatic not to pursue the subject any further. "Thanks for the report," I offered as humbly as possible, and then added, "See you later." There seemed to be more spring in my step as I skipped home than when I had crossed the street to visit with John.

Are We Having a Christmas Goose?

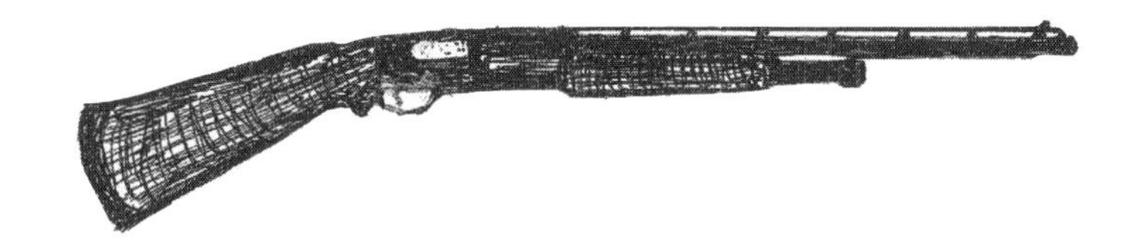

Christmas Eve and I had not yet started to prepare my part of tomorrow's dinner. Our home had become the gathering place for several faculty families who did not have relatives nearby to spend the holiday. All fall I had planned to hunt geese at Beaver Creek Reservoir in the foothills of the Shenandoah Mountains but something always came up that interfered—a deadline to get a psychology paper written and submitted to a journal; caulking a leaky chimney seam on our new house; deer hunting with my old friend, Tom, whose wife wouldn't let him hunt alone since his heart attack two years ago. As cold and windy as the weather was today, I wondered if I would be able to supply the main dinner course for my family and friends who planned to get together tomorrow.

I had scouted two dozen Canada Geese that hung around the reservoir during the summer and fall. They were big wild birds that wouldn't let me get within 500 feet of them when I fished the lake from my canoe. On a blustery afternoon, like today, they would be hunkering in the sheltered part of the reservoir where Beaver Creek entered the lake if they hadn't migrated south when the cold front descended last night. The upper reaches of the reservoir were surrounded by protective banks of hardwood trees on both sides. I would have to ride the canoe a couple miles, leave it behind on the last shoreline bend, and sneak through forest a half mile to get within shooting range.

I reviewed a myriad of thoughts as I lashed the homemade wood and fiberglass canoe to the top of my International Scout. *"It's so windy it will be dangerous in the canoe on the water. What if I tip over? I've been married five years and we are expecting our second child next month. Marilyn wouldn't like to be widowed so young. We could eat venison for Christmas dinner instead."*

"We have never served a Christmas goose before," I pondered wistfully. I was intent to roast a goose for dinner tomorrow. *"I'll stick close to shore so I can swim to safety if the canoe upsets."* That settled it.

"Don't worry, I'll be back before dark," I hailed as I shut the back door after saying goodbye to Marilyn and our three-year old daughter, Shelby.

Fifteen minutes later I parked the Scout on a grassy incline by the reservoir dam and unloaded the canoe. Carefully, I laid the barrel of my Winchester 12 gauge on the thwart ahead of where I would be sitting and stashed a five-gallon bucket filled with rocks in the bow of the canoe so the front end wouldn't stick out of the water and catch the wind when I sat down in the rear.

I headed directly into the fierce west wind. It took all my strength to paddle into the gale. Sometimes I felt precipitously close to upsetting when the wind struck the 17-foot watercraft broadside while I followed the curved part of the shoreline. I was glad to finally reach the quieter cove where I could park the canoe to stalk the geese on the ground.

I glanced at my watch. *"Four o'clock already. I hope the geese are where they are supposed to be."*

Making good time, I scrambled westward through the woods behind a hill until I had to slither along on my belly as I neared the top. I crept behind the thick trunk of a leafless yellow poplar between the water and me and cautiously peered around the bark toward the reservoir's upper end.

"Yes, they're there! But they're a hundred yards away. That's too long a shot for my 12 gauge, even with a full choke and buckshot. I need to get closer."

I surveyed the slope between the tulip poplar and the lakeshore. *"Little cover to hide behind. Only a couple azalea bushes twenty yards ahead. The leaves are gone from all the other shrubs."*

As I lay on the ground, the temperature of the cold earth penetrated through my coat. I wished I had brought gloves—the metal gun was icy. But a bigger dilemma was at hand. *"How can I get closer to the geese for a better shot?"*

Pondering the situation for several minutes I decided I would slide, ever so slowly, on my backside down the slope toward the water. I hoped the geese wouldn't spook as long as I looked like a wounded critter that couldn't move fast and wasn't a threat. I would have to hide my gun under my camouflage coat and keep my hunting cap over my face as best I could.

I readied myself and purposefully left the front of my coat unzipped so I could quickly bring my gun into shooting position if need be. I squiggled from behind the poplar tree and laboriously propelled myself downhill, using my legs to pull myself along and dragging leaves and dirt with my butt.

With cocked heads jerking, several geese surveyed me as they swam in the open water directly ahead of me. Two honked softly, not the loud alarm call, but a, *"heads up—what is this strange thing?"* sort of call. Over the next several minutes I galumphed a good 50 yards down the slope. The geese strayed as far toward the opposite shore as they could without leaving the water. Now we were 60 yards apart. *"Another push or two and I might be close enough for a shot."*

Crack! A twig snapped under my weight. It was enough to trigger alarm. I had reached the geese's fight or flight distance. Several barked loudly and flapped hard to rise from the water's surface.

I rose quickly, drew the gun from under my coat to my shoulder, aimed at the closest bird winging skyward and fired.

Feathers flew but the bird flapped its wings more desperately. I fired again. The bird fell into the water and swam vigorously toward the opposite shore as I raced downhill for a closer shot. We reached shorelines simultaneously and I fired again. More feathers flew. However, my gun, with its required plug, was out of shells and the goose was climbing the next hillside away from the lake. *"How can I get closer? The bird can't fly so I should be able to retrieve it if I can get over there. But it's getting too late to go for the canoe and will be dark by the time I get back here."*

"Maybe I can cross the creek above where it runs into the reservoir." I could see the inlet. I ran along the shore until I reached the confluence. Here the stream was only about twelve feet across but maybe too deep and flowing so swiftly that crossing on foot would be difficult.

"If I leave my gun here I might be able to jump across," I contemplated. I took a few steps back, laid my gun on tufts of still green sod, pulled off my coat, got a run and leaped as far as I could. My right foot landed on the bluegrass covered bank and the other sank down two feet into water but my momentum kept me leaning forward and I was able to pull my wet leg onto solid ground.

I ran to where I last saw the goose scurrying and began a search pattern back and forth—I scoured fallen tree branches, azalea clumps, and weed patches. I worked my way up the hillside, but with no luck I returned to the lakeshore and scanned the horizon. I couldn't see the goose swimming anywhere.

I made my search pattern bigger and worked upside the hill again. Daylight was dwindling. Sodden gray clouds were hurrying nightfall along. Sweating, I stopped walking and thought for a while. *"Where would the goose head? Of course, it would try to get as far away from me as possible."*

Looking around, I spied a woven wire fence with a couple barbed wires above it stretching through the trees along the hilltop. Years ago farmers let hogs run wild in the woods on their property and the wire fences marked off the boundaries.

Quickly I jogged to the fence. Just as I reached the wires a gray and black form rose out of a pile of leaves and faced me down. The goose hissed as it raised its wings. I could see blood on the underside of its right wing. I grabbed at the goose, but it wheeled and turned toward the lake.

I didn't know geese could run so fast. Each time I attempted to grab an outstretched wing or leg, it evaded my hand and I had to regain my course as it set off in a different direction. We neared the water's edge.

Just as the goose leaped into the water I threw myself on top of it. Never mind that I would get wet. I already was drenched in sweat as well as having a water-soaked leg. I got hold of the goose's neck and broke it in a swift twist. I righted myself and tossed the goose onto the shore.

Gradually my huffing and puffing subsided as the goose's death throes diminished. "*What a huge bird!*" I picked up the goose and made my way back to the stream where I had crossed earlier. Not wanting to lose the bird again if I tried throwing it across and missed, I took a good run, jumped and landed a couple feet from the other side of the creek, goose in hand. The water was shallow though and only my right leg became soaked up my knee, but so what! It had been wet already.

Gathering my coat and gun, I hurriedly retraced my path toward where I left the canoe. There were no shadows and daylight was spent when I reached the cove and loaded the canoe. Fortunately the wind had died down as evening set in and the remaining breeze helped push me back to my vehicle. The sky was pitch-black when I reached the parking area and snow flurries were swirling around me.

Twenty minutes later as I motored up our driveway, all the lights were on in the house and the outside yard was lit up as well. I could hear Christmas music emanating from the house, even through the rolled-up windows of my Scout. Marilyn liked it that way, music shaking the rafters. I honked the horn to let Marilyn and Shelby know I was home.

The front door of our newly built country house opened widely. Shelby waved. A lusty three-year old voice cried out, "Are we having a Christmas goose?"

"You bet," I answered. It was the best goose I've ever eaten.

Manteno

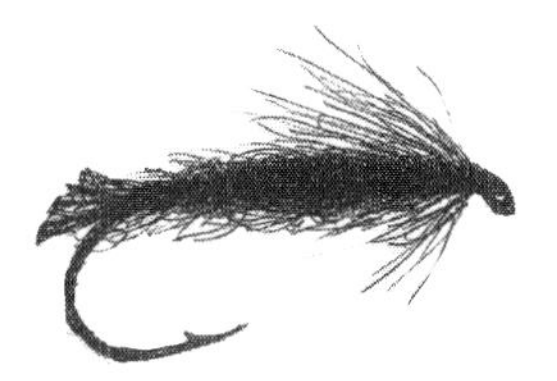

Manteno Reservoir in western Iowa was like Beaver Creek Reservoir in Virginia: a training facility and the source of great fishing and fun for the whole family. My family always found Manteno Park to be a relatively underused camping site, replete with mature trees, rugged hills, wildlife, and opportunities for joyful commune with nature. Taking a break from farming and professional life to camp at Manteno Park renewed our energy levels.

When Shelby and Jon were youngsters we often played "*Three Billy Goats Gruff*" on the footbridge over a ravine leading to Manteno Reservoir. I hid under the wooden structure while Marilyn, Jon, and Shelby cautiously tiptoed across the planks. As they approached the middle of the bridge, the wicked troll snatched at their heels. Shelby and Jon shrieked with delight as they scrambled across the bridge and out of the troll's grasp.

One warm, June, Saturday evening, Marilyn, the kids, and I decided it was time for a respite. The field work was finished for a while, the family needed a break, and Manteno provided the amenities we needed. We loaded the Coachman camper with our supplies on the pickup truck and set off for the park. After our perfunctory sojourn to the *"Three Billy Goats Gruff"* bridge, we staked out our campsite. That no one was camping nearby made our spot all the more pleasant.

The star-filled evening around a blazing campfire with hotdogs and marshmallows brought the contentment we craved. We huddled together as I played the guitar and we sang every song the kids knew.

Shortly before daybreak, while everyone else was still tucked in their beds, I arose and quietly donned my waders and flippers. With fly rod and float tube in hand I stepped to the reservoir near to where we had parked. Already vehicles with trailers and boats were pulling into the parking lot across the lake. I wondered what was going on.

As I paddled into the lake and started casting, boats of all sizes and values whizzed by, heading for various remote bays and weed beds along the shore line. At first I paid no attention to the passersby and went about my business of casting my favorite little black fly towards the shore. Within minutes I landed nine keeper bluegills and a 16-inch bass. I noticed a couple boats that flashed by were now slowly approaching me. Every boat had two people in it. One blue and white boat contained a fisherman who was aiming binoculars at me.

Ignoring their curious glances I continued casting toward the shoreline. Within a few minutes I caught three more nice sized bluegills, one crappie and another 16-inch bass. The boat with the man who had trained his binoculars on me drifted quietly to within 50 feet. Both persons on the flashy blue and white bass-boat hurled big lures 15 feet on each side of my fly, so I paddled away a couple hundred feet.

Here the shoreline contained no structure except for a single big elm that had fallen into the lake. The massive trunk and a few branches stretched 40 feet into the reservoir. I cast my little black fly into the V-shaped water where the log angled from the shore. A big ripple began emanating outward and I lifted my rod expectantly. I yanked too fast, there was no resistance on my line. Quickly I looked around to see if any observant fellow fishermen were watching, but no one seemed to be facing my direction at the moment. As surreptitiously as possible I kicked my way down-shore to the next inlet.

Within minutes, the blue and white boat followed me into the inlet. I got in a few more casts before the fellow fishermen began throwing their lures over my fly. I managed to land two ten-inch crappie before retreating again.

I decided to head back to the downed elm and to try again for the fish I had missed earlier. As I moved up-shore and approached the big log wallowing into the reservoir, I tossed my little black fly 50 feet into the corner that the log made with the shore. The fly gently settled onto the surface. I tugged lightly on the line. Big commotion! Water flew in all directions. The fly line went taut and I jerked the fly rod upward but the tip stayed pointed into the water. I tried to retrieve the fly but the line didn't yield. Fishing line started to unravel from the reel. I could see the water stir under the log, so I applied more pressure to keep the big fish from wrapping itself around a branch. A huge black bass managed to throw his bulky body out of the water and landed with a loud "plop!"

Furiously I cranked on my reel. The unruly bass begrudgingly allowed me to drag it toward me but I lost nearly all the line I gained whenever the lunker lunged toward the submerged log. Luckily, I held the bass from hunkering in the submerged tree branches and eventually headed the massive fish into open water toward me. By now two boats were streaming toward me, one on my left from the inlet I had vacated and another from my right.

As the bass' tussles with me gradually weakened, I maneuvered the big brute into my outstretched fishing net. Raising the net above the water, thankfully I said, "Yes!" just loud enough for the neighboring fishermen to discern.

"Get off the lake!" a man yelled from my right side.

As I hooked the huge bass onto my stringer, the blue and white boat coursed directly in front of me.

I paddled backwards but I could see that the way toward my campsite, 200 yards to the right, was blocked. Changing course, I paddled toward the distant shoreline next to the dock where the many boats on the reservoir had originated earlier in the morning. As I approached shoreline, a man wearing a name tag around his neck and holding a clipboard in his left hand walked toward me as I was backing onto the beach.

"What number are you?" the man queried.

"What do you mean?" I replied.

"Aren't you registered in the tournament? Do you have a number?" Then, spying the stringer containing crappie and bluegills as well as bass, the official looking man suggested, "Only the bass will count."

"No," I responded. "I didn't know there was a fishing tournament going on."

"Sorry," the official-looking man quietly uttered. Then he said, "Let me see your fish."

I raised the stringer of fish that was mostly submerged next to my float tube while still standing in a foot of water at the shoreline. With difficulty I raised a heavy chain stringer with eighteen fish, the largest of which was the huge-bellied bass on the last section of the stringer.

"Wow, you got at least 12 pounds of bass and that last one's at least 6-1/2 pounds," the tournament official offered. The other two bass are okay but we don't count the crappie and gills 'cause only bass count in this tournament. I think you would win this tournament easily if you had registered."

I asked the official why one of the competitors told me to get off the lake. The official said, "He probably thought fly fishing for bass isn't allowed, but there's nothin' that says you can't use flies."

At that, I decided it would be best to get out of the water here and to walk along the shoreline the long way around the lake to our campsite. As I made my way down the trail, I passed under a banner stretched across the trail that proclaimed, *"Happy Hookers Bass Tournament."*

Years passed. Now some time after my unofficial tournament victory, eleven-year-old Jon and I decided over Sunday brunch that Lake Manteno might be a likely spot to open the spring fishing season. Although it was still late March, the warm sun had heated the outside air to 60 degrees. Good fishing weather!

Jon and I loaded our homemade canoe onto the farm pickup truck and headed toward Manteno Reservoir. By one o'clock we were on the water. Jon wore a short sleeved shirt in the warm sun and I wore only a fishing vest over my t-shirt. We paddled the canoe toward the far end of Manteno Reservoir, hoping that crappie would be hungry. Ice still extended a ways out from the shore.

As we paddled toward the dam at the deep end of the reservoir, we noticed another boat halfway toward the far end. When we floated past the aluminum craft, we exchanged waves. I guided us around the dog-leg toward the far shore. "Shouldn't we try that bay, Dad?" Jon suggested as he pointed to an inlet to our left. "We usually catch them back there."

"Let's see if the crappies are biting by the dam first," I responded. "Why don't you go ahead and start fishing."

Jon unhooked his little black fly from its eyelet on the bottom-side of his rod and cast it toward the dam as the canoe gradually slowed. Hardly had the loose fly line straightened out when Jon's rod jerked sharply downward. Jon reeled vigorously and soon landed a hefty 11-inch black and white speckled crappie. I let the canoe drift and cast into the open water toward the dam as well. My little black fly sank several feet when something yanked hard and I set the hook. A few seconds later I brought up a decent sized, salt and pepper colored crappie as well.

Over the next couple hours, Jon and I were in fishing heaven. Thinking we had caught enough fish and because a gray cloud bank was easing eastward toward the mid-afternoon sun, we decided to head toward the dock where we had put in our canoe.

As we paddled around the bend in the reservoir, Jon and I approached the anchored, flat-bottom aluminum boat we had passed earlier. Its occupant yelled, "Hey Mike, how the hell are you doin?"

"Fair to middlin," I replied, as we braked the canoe to stop near the boat.

"Did you catch anything?" the suntanned, unshaven but pleasant man in the boat questioned.

"Oh, a few," I replied.

"Let's see 'em," the fellow fisherman requested.

I recognized our friendly fellow fisherman. It was Larry, a former neighbor who, as well as his brothers and father, liked to fish.

I elevated a five-gallon bucket from the bottom of the canoe just ahead of my feet and tipped it so Larry could peer into the bucket. It was three quarters full of fish.

"Holy shit," Larry uttered. "What did you catch them on?"

"Oh, a little black fly I tied," I replied.

"The hell," Larry gasped. "I've been fishing all afternoon and I only caught a couple."

With a twinkle in my eye, I called to Jon in the front of the canoe, "Hey Jon, show Larry your bucket."

Reaching behind him, Jon couldn't raise the heavy bucket but tipped it just enough so that Larry could see that it also was nearly full of fish.

"Holy xxx!!**xx!!!**!"

"Time to go home, Jon," I directed.

Quite a few years later, Jon, Walt, and I launched our canoe from the oft-used dock at Manteno Reservoir. It was a warm sunny Sunday afternoon in June. Jon sat in the front of the canoe, I sat in the back, and

Walt sat on a fishing tackle box in the middle of the canoe. Walt wore a green lifejacket under his tattered fishing vest.

Walt had endured a series of strokes over the past several years that affected his body's left side: he walked with a severe limp but he could still cast a fly with his right hand, his dominant hand.

Happily, we pushed away from the dock of Manteno Reservoir. Even Walt, who lived with Michi in Idaho, was familiar with Manteno Reservoir. We hoped to catch a good bunch of fish before Marilyn and Michi arrived with a supper that we planned to eat at the picnic site nearby. The afternoon passed quickly and Walt caught as many fish as Jon and I. We nearly filled two five-gallon buckets with crappie, bluegills, and bass. It was time to enjoy supper before we went home to clean our fish. I hoped to send some fish home with Walt and Michi the next day when they were scheduled to take the airline back to Idaho. The Rosmann freezer would be replenished for several meals of fish as well. Unfortunately for Shelby, she was away at college.

Marilyn and Grandma Michi arrived in our family car just as Jon, Walt, and I eased up to the dock. After exchanging greetings and congratulations, we sat down to a sumptuous picnic supper of tempura fish, sweet potatoes and peppers, sushi, cucumber salad, pickled burdock, and mochi for dessert.

When supper was finished, it was 5:30 PM but the day was still warm. Michi asked if Walt wanted to ride with Marilyn and her back to the Rosmann homestead but Walt said he wanted to ride with Jon and me.

With difficulty, Walt was able to drag his weak left leg and arm into the front seat of the truck, where he sat contentedly and placidly with Jon on his right side and me on his left. Marilyn and Michi had already headed homeward. Just as I backed away from the parking site I heard

Walt gasp beside me and stiffen. Walt's head slumped forward and his body trembled. He stopped breathing.

I slammed on the brakes and asked Jon to run to the Manteno Park caretaker's cabin to call for help. As Jon leaped from the pickup and headed down the gravel path, I yelled out the window, "Wait. We can get to the hospital faster ourselves than to wait for the ambulance."

Jon raced back to the pickup truck and jumped in beside his grandfather. Walt's face was ashen and still. "Walt, are you okay?" I asked hopefully.

No response. Jon put his hand on Walt's right arm but Walt did not respond with any sensation.

Jon extended his left arm over Walt's chest to keep him from slumping onto the dashboard as the pickup truck lurched downhill toward the highway. As the vehicle reached the first stop sign a half mile away, Jon and I heard Walt gurgle. He seemed to be trying to get his breath. Then Walt sucked in another whiff of air.

I stepped on the accelerator to race toward the closest hospital 20 miles away.

Ten miles down the road Walt twitched and seemed to breathe somewhat easier. I decided to drive past our farmstead, which was only a little out of the way to the hospital. As the pickup truck neared our home, Walt seemed to mumble something. I wanted to warn Marilyn and Michi about what was happening. When we pulled up in front of the house I honked the horn. Marilyn came running out of the front door.

"What's wrong?" Marilyn asked nervously.

"Something's wrong with your dad. He might have had another stroke," I blurted.

Just then, Walt mumbled, "I'm okay."

With much uncertainty and doubt, Marilyn and I looked at each other. We loved Walt greatly and wanted to do what was right for him.

"It was just another seizure," Walt quietly uttered.

Hearing her husband's comment as she was coming out the front door of our farm house, next to the canoe laden pickup truck, Michi said, "He has them all the time ever since he had his last stroke."

With much relief, Jon and I maneuvered Walt out of the truck and onto the couch in the living room. Jon and I talked a good deal about the event as we cleaned the two buckets of fish in the basement afterwards. That night as Marilyn and I were lying in bed, we wondered why Walt and Michi had not told us that Walt now was having seizures.

When I asked Walt the next morning, he told me, "I didn't want to miss going fishing at Manteno."

One Shot

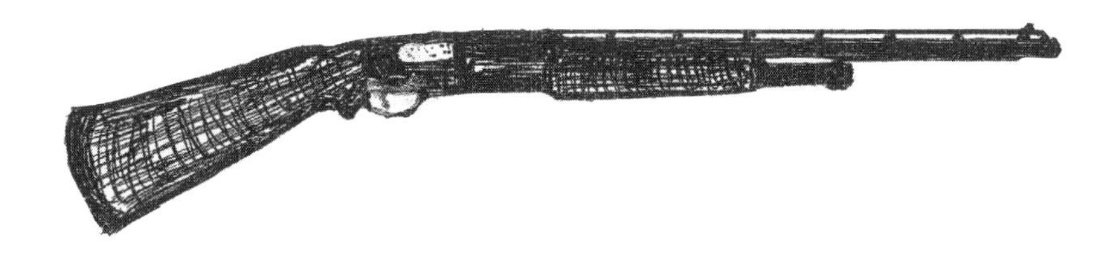

TGIF! The mid-September work week was over and I was tired. A lot of school kids were being referred to Prairie Rose Mental Health Center, the outpatient clinic I directed in Harlan, Iowa. Apparently, teachers didn't want to have to put up with the barrage of problems that the parents and kids hadn't addressed during the summer, such as kids' refusing to follow directions. Now it seemed as if everyone was in a "fix-it" mode.

As I walked to my car, in the parking lot outside the mental health center, I noticed that the air had a fall-like chill. "*Tomorrow is the opening day of duck season,*" I remembered. I decided to drive past a farm pond on the way home, six miles from town, to see if there were any ducks on the water. As I drove past the farm pond I could see the shallow body of water had silted in considerably during the summer and now cattails and duckweed covered nearly all the remaining surface of the water. A flock of teal crowded into a small patch of open water. As soon as I arrived home, I called the owner to ask permission to hunt there tomorrow. Farm chores seemed to go easily that evening because the pond owner told me it was fine if I hunted there. I figured that the teal would hang around the area a day or two.

That night I oiled my 12 gauge, pump action, Winchester and screwed in the, "*Improved Cylinder*" choke. I plotted strategy for tomorrow's hunt. I would try to sneak through the marsh grass as close as possible to the pond and jump-shoot the ducks when they flushed. Hopefully I would be able to knock down one or two of them before the entire flock escaped. Then I could set out a spread of decoys and hope that other flocks of migrating ducks would happen to fly over the little pond. Already the cares of a heavy workload were receding into the background and I was gaining the kind of perspective that hunting, fishing, and farming usually yielded.

I liked to think I was a better psychologist because of my regular encounters with nature. One time I took a client fishing because he was so depressed he couldn't say much of anything and by the end of the day he was talking animatedly. And I didn't charge him for the session!

By mid-Saturday morning I headed to the pond. I parked my pick-up truck a quarter mile away on the gravel road leading to the pond. Grabbing my shotgun, I crawled through the fence and carefully threaded my way through the rows of tall corn to reach the marsh that surrounded the pond. As I neared the edge of the cornfield, I crawled on my hands and knees. "*Would the ducks still be on the pond?*" I wondered.

Reaching the marsh grass, I cautiously parted the strands of reed canary grass to peer toward the pond twenty-five yards ahead. Lo and behold, the ducks were still there. I pressed off the safety of my shotgun, hoping the click wouldn't make too much noise. I made sure that I had my duck call in my left pant pocket. I had left the decoys back at the truck, but I could always go after them later to set up a spread.

"*Now is the time,*" I thought. I stood up. The flock of teal immediately burst skyward in panic. When they arose thirty feet over the pond, I aimed my shotgun at a drake in the center of the hoard. I squeezed the trigger. The shotgun blasted and the recoil felt good. Quickly I pumped out the emptied shell and brought the next shell into the shooting chamber.

I couldn't believe my eyes! The sky seemed to be raining ducks. They fell everywhere onto the pond. I pressed the safety into the off position and quickly surveyed the carcasses rippling on the surface of the pond and one or two birds still fluttering in their death throes while the remaining teal were hightailing southward. I shoved my way through the reed canary grass to the water's edge.

"*How will I retrieve them?*" I pondered.

Quickly I unlaced my shoes, kicked them off and pulled my pants off. Wading in my underwear into the shallow pond, I collected all the teal I could find. It took two trips to locate all the dead ducks. I counted the pile of carcasses next to my shoes. "*Nine ducks, all in one shot!*" I exclaimed to myself.

I was so excited I didn't seem to notice that my socks, pants, and shoes became wet when I pulled them on. Then a troubling thought crossed my mind. "*The limit is six. What will happen if the game warden checks me?*"

"*Well,*" I philosophized with a smile, "*If the warden checks me, I've got my license. I'll tell the truth. I fired only one shot!*"

The Meaning of Christmas

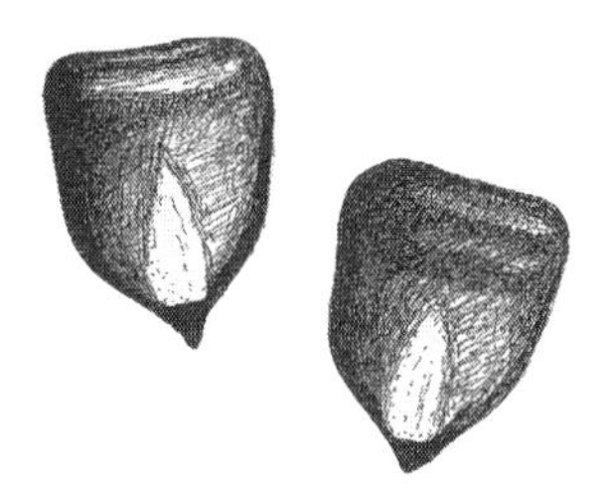

I believe my generation is the last in rural America in which the needs of the family farm took priority over the desires of the children and, "what is good for their development." Consequently, I learned to arise before 6 AM at the beckoning of my father's call, to milk cows by hand and to feed my 4-H project cattle. Nowadays, most farm children are more concerned about getting their school books and playground games organized before the arrival of the school bus than completing the farm chores. I don't know who is more right, my parents who stressed hard work, or my generation who emphasize, "doing everything right," for our children. The Christmas season epitomizes the disparity of values with which I was raised and those that my wife, Mariiyn, and I attempt to apply in our child-rearing. Yet, as much as times and priorities have changed, the real meaning of Christmas continues to flourish at least one more generation in our family.

Perhaps our children will raise their sons and daughters differently also. Will they succeed in imparting that, "giving is more honorable than getting?" Will their children be able to experience horses that nuzzle on their hands, cats and dogs that follow them from barn to barn as they tend to the needs of cows and pigs, asparagus and raspberry patches, and a tree house in the farmstead grove? What will they hold to be most dear about Christmas and which Christmas will be most memorable?

I cannot honestly say what it is about Christmas that I hold most dear. To be sure, I have been thrilled at my share of midnight masses in our little German-Catholic community. When the snow lightly settled on the silent countryside while my family journeyed to midnight church services, I felt a tingling in my heart. When the choir of mostly farm men and women broke into four-part harmony while singing *Silent Night* I felt a bit of a tear enter my eye and a chill down my spine. When Shelby, our 14-year-old and first born child, was just five days old for her first Christmas in the arms of her happy mother with proud Grandpa and Grandma nearby, I felt I was the luckiest man in the world. How does one top those moments of ecstasy in deciding which Christmas is most precious?

After many days of contemplation I still cannot say with complete certainty which Christmas is most memorable. But I have arrived at what I feel most deeply about Christmas. I truly love getting up in the gray of dawn and brewing a steaming hot pot of coffee to sip while going about morning chores. Christmas morning stirs in me a special affection for my cattle and prompts me to drop extra shovels of sweet corn and layers of aromatic alfalfa bales into the bunks of the powerful herd bulls and gentle cows. The rooster pheasants beating their wings and crowing to the harem of hens still in the spruce and pine windbreak tells me that this rich farmland produces bounty for wild and domestic alike.

An hour later as I approach the farmhouse, now with several lighted windows, I can hear excited shouting even though both the storm door and heavy wooden inside door are latched, as the kids discover unpredicted treasures in their Christmas stockings. They run to the entryway as they hear me kick off my boots. Jon hugs my coated waist, oblivious of the chaff brushing onto his pajamas, telling me, "Thanks Dad" for the newfound Nintendo game.

Shelby plants a shy, adolescent kiss on my frosty cheek as she says, "Thank you," for the new hair dryer which mom so thoughtfully remembered.

Then Marilyn hustles to the doorway, throws her arms around my neck, and says, for the sixteenth consecutive year, "This is the best Christmas yet." I forget all my inner debates about what values to impart to our children. I utter an unspoken prayer of thanks to God and hurriedly strip off my heavy outer clothes to investigate what might be inside the neatly wrapped packages tucked inside my long red stocking by the fireplace.

Winter Breaks

The first Monday of February in Iowa emerged as a bright sunny day. All the cattle had been fed, another batch of ear corn had been ground for the next five days and no chores needed to be done just now. Time to go fishing. Taking a winter break would be more difficult a month from now when the heifers were calving and would need checking every two hours day and night.

Dan's pickup truck pulled into the yard with Monty and Steve beside him. When Dan called yesterday to ask if I wanted to go fishing with them, I said, "Of course." I volunteered to supply the sandwiches and bait if Dan would furnish his hand cranked ice auger. Monty and Steve would supply the drinks.

I stowed a lunchbox packed with sandwiches and a half dozen short ice fishing rods in the corner of the truck box. I threw several five-gallon buckets in the truck box for good measure so we would have something to sit on and ample containers to store the many crappie, perch, and bluegills we expected to catch.

When I asked if I should drive my own truck because there were four of us and only one seat, Dan said we could all squeeze into his truck. Monty insisted that he always liked to sit close to Dan anyhow. Dan commented that this was fine as long as Monty didn't tickle him. Steve promised that he wouldn't belch or break wind so I packed in next to Steve.

Twenty minutes later four farmers gladly piled out of the truck at Mush Anderson's pond, uttering, "Oofta-tooftas" because Steve had broken both promises. Steve volunteered that he would fish downwind from us but nobody believed him.

Taking turns when our arms got tired, we quickly drilled a dozen holes with the auger through eight-inch-thick ice on Mush's pond and dropped minnows and grubs tied to midget ice fishing rods into the holes. Time for cigars and lunch.

Lunch consisted of cigars and *Bud Light*. Nobody wanted a sandwich. No one had any fish yet either. We decided to experiment with various holes interspersed throughout the pond to attempt to locate where fish might be lurking. We knew Mush's pond held decent-sized fish because Dan and I caught a nice mess last summer.

By midafternoon, still no fish! Dan and I headed up the frozen stream leading into the pond. Many willow, ash, and box elder trees lined the creek banks and stumps of dead trees protruded from the streambed. This stretch held a lot of crappie last summer, so maybe it would be worthwhile trying here again. Meanwhile, Steve and Monty tended the many open ice holes on the main part of the pond.

I had just started to drill a hole next to a protruding stump when I noticed water seeping onto the ice around the tree stump. As I started to raise the auger I realized that probably the ice was thinner here because the fresh water entering the pond was warmer than the water in the main body of the pond. Too late! The ice broke beneath me and I sank over my head. When I bobbed up again I thrust the ice auger over surrounding firmer ice and held on. Dan was coming to my rescue, but I told him to stay back because he would break through as well.

I kicked my way toward the nearby tree trunk, using the auger to help bear my weight. Fortunately, the tree trunk angled sideways 45 degrees so I was able to hang onto the trunk and gradually sidestep up the trunk to the surface. I managed to get all my body out of the icy water. As I hung onto the trunk with one arm I swung my body toward the shoreline just a few feet away. I landed with a thunk and broke through the ice but the water here was only a few inches deep. Gathering myself and the ice auger, I hastened back to join Dan, who said he would go ahead and warm up the pickup truck so I wouldn't become hypothermic. We could hear Monty and Steve laughing and yelling in the distance.

As we drew closer to Steve and Monty, we spotted a couple dozen bluegills and crappie that were thrown on top of the ice near the holes where Steve and Monty were busily scurrying back and forth. As fast as they could reel in a fish, another would bite.

"Maybe we should fish for a while," I suggested to Dan. Although I was wet from head to toe, I wasn't cold. Walking vigorously had warmed me up and the icy, 10 degree wind had frozen the outside shell of my garments to create a windproof barrier. Dan loaned me his stocking cap while he pulled the hood of his coat over his own head.

For the next hour, four happy farmers scrambled to keep up with all the bobbers and tip-up flags popping up and down. The fast fishing action and adrenaline kept me warm. When we headed home to carry out evening chores, each of us had a five-gallon bucket nearly full of fish.

Years earlier I had learned from Walt that midwinter fishing could be some of the most enjoyable and successful piscatorial pursuits. Marilyn and I often drove to Burley, Idaho, from Salt Lake City to take a break from our responsibilities. Marilyn was a faculty member on the nursing staff at Weber State University and I was finishing up my psychology training. At least twice each winter, Walt and I would visit the Thousand Springs area along the Snake River. We fished the pools below the myriad of springs that gushed forth from the rocky northern wall of the Snake River channel. Most of the rainbow trout we caught were nine-twelve inchers that vigorously attacked tiny midges, sizes 22-26, Walt and I tied. Even if fishing was slow at Thousand Springs, it was still a lot of fun. There seldom was any wind in the riverbed protected by sheer 200-foot high walls on either side of the river. The watercress was especially fresh and sweet in the pools below the springs where the water was a constant 45°F as it emerged from the slate colored rock. The Thousand Springs' water was said to have taken several years to flow underground from a large desert depression surrounding Craters of the Moon National Monument in east central Idaho. Surprisingly, Walt and I usually were the only Saturday venturers during these winter escapes to Thousand Springs.

Winter didn't deter Walt and me from fishing any water that was ice-free. There also were occasions when I fished with midges by myself on Utah streams such as the Weber and Duchesne rivers. The tiny midges could be very effective, even on trout as large as 18 inches if you were careful not to horse them in too quickly. But, there was no winter fishing as good as a little ranch reservoir in southern Idaho where Walt and his friend, Bert, introduced me to midwinter crappie fishing.

After I had purchased my own float tube, Walt and I frequently relied on the excuse that we needed to replenish our freezers and those of several families who had come to depend on us to supply them with fresh fish. Generally we used streamer flies.

Even though it had snowed a foot the night before, Walt and I set out in his '59 Chevy Impala to pay a visit to the little, sulfurous, spring-fed reservoir. Walt quickly drove the 75 miles of cleared highway on a late Saturday morning. The nine-mile gravel road to the reservoir also was open, but the trail to the dock on the east side of the reservoir was unplowed. Walt said, "Let's go in from the west side instead."

I cautioned, "Do you think we should try this?" The snow was mounded into four-foot deep drifts on parts of the trail.

"No problem," Walt replied as he blasted through the first big snowdrift. He gunned the Chevy to break through a couple more drifts and we were soon halfway down the trail to the reservoir. A long stretch of heavy snowpack piled next to sage and rabbit brush lay ahead. We plunged through the first 100 feet, and then Walt's Chevy ground to a halt, stuck in the deep snow and high centered. We were at least 100 feet from the end of the drift.

"We can dig our way out," Walt suggested as he handed me a scoop shovel from the trunk and as he grabbed a long-handled spade. We dug and sweated to remove the snow under the car until it reached firm footing and we scooped out a trail ahead of the car. By the time we were done, it was already 3:30 PM.

Fearing that we might not have enough daylight to get back home, I suggested that we turn around now.

Walt said, "No, let's go fishing."

Walt busted through several smaller snowdrifts and finally we arrived at the bank of the reservoir. After donning insulated waders and tucking into float tubes, we paddled into the open waters of the spring-fed channel.

Two hours later and with a full moon beaming overhead, Walt and I got out of the water, each with 40-50 crappies. The extra weight of the fish and the successful exploit must have inspired Walt, for he charged his Chevy down the trail that we had forged earlier. Snow flew in all directions and we nearly came to a stop a couple of times, but always the car kept moving ahead. Finally, we reached the gravel road. We even managed to arrive back in Burley by 10 PM, after supper at Emma's Highway Cafe.

It was after midnight when Walt and I finished cleaning the fish and I, completely bushed, joined my wife in bed. Marilyn asked, "Did you enjoy the break?"

"I certainly didn't think about studying or work," I responded.

"Good," Marilyn sleepily commented.

Three Cigar Day

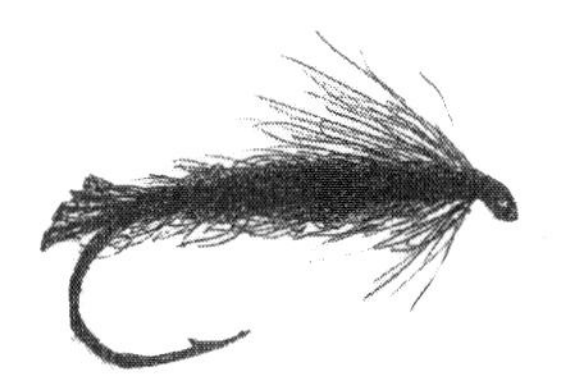

Usually two cigars will last through most days of fishing. I always contended that the smell of tobacco on my fingers made the flies more natural and earthy smelling, but this was my personal observation and not a scientific finding.

I had arisen at 6 AM on a fine mid-October morning. Marilyn and I were visiting her parents in Burley, Idaho. Marilyn's father, Walt, had suffered another stroke several weeks earlier and now could walk only with great difficulty.

I treasured these trips because Walt and I would swap stories and I could pick up useful hints to improve my fly fishing techniques.

Walt was a master fly fisherman but he never let you know that directly. I was certainly less shameless about reporting my successful fishing and hunting exploits, but I had learned a few things from Walt, such as not telling anyone who couldn't be trusted where the big ones could be caught.

Last night Walt shared more stories about the river to which I was heading today. Walt and I had often fished there but I had not exhausted my father-in-law's supply of lore about this magical river. The river flowed through central Idaho lava beds. Steep canyon walls had to be scaled in order to reach the riverbed 150 feet below. Climbing up the rugged trail after a hard day of fishing was a taxing enterprise, made all the harder with a hefty stringer of fish. A special strain of rainbow trout flourished in the cold waters that emanated from a dam several miles upstream.

The river nearly lost its fish population when the local power company shut the water off at the dam to make repairs two summers ago. That fall fishing was terrible and I didn't even get a nibble. Last year when I tested the waters I worked hard all day to catch two respectable 18-inchers.

Back in the glory days, before the repairs, fishing this river was a mystical experience. Walt caught his largest trout here on a fly on December 24, 1972. I helped Walt land the fish by loaning him my fishing net, because Walt's net wasn't deep enough to hold the monstrous rainbow. It ended up at 29-½ inches long and weighed 10-½ pounds. I dreamed of catching such a wonderful trout someday. Maybe today—but then, I always thought this when I headed to these relatively unknown but glorious waters. I had managed to catch a 5-½ pound rainbow here five years ago. Each fall, after the crop irrigating season ended, the water master cut back the flow of water coursing through the dam to just enough to sustain the fish population downstream. Rainbow trout congregated in the deep holes, including huge lunkers. Walt had invented a fly called the "Big Shrimp Special." It imitated the millions upon millions of fresh water shrimp that swarmed in the watercress and moss that lined the banks and much of the streambed. I had learned to tie the Big Shrimp Special and wanted to compare my creations with Walt's inventions today.

As I drove over the dusty desert road, to the banks of the river, I could see that I had the river all to myself when I arrived. It was an outstanding day, not a cloud in the sky and the frost had already melted in the first rays of sunlight. I parked my Jeep about ten feet from the cliff's edge and peered into the water below. "*Great,*" I thought as I could see that the irrigation flow had been turned off. I would be able to wade anywhere I wanted. I could hardly contain myself as I pulled on my waders and readied my gear. In my eagerness to fish, I made the same dumb mistake I usually make when I'm excited: I forgot to thread my line through one of the eyelets on my fly rod and didn't notice the mistake until I had already tied on a fly. "*Dang,*" I thought, as I snipped the fly off and rethreaded the line properly. I was losing precious time. "*Slow down,*" I admonished myself. One more sip of coffee and I was off.

Cautiously I stepped down the steep trail and pushed aside the many still-green willow shrubs until I reached the water's edge. I had tied on a Big Shrimp Special that I concocted on Walt's fly tying bench last night. I cast into the first pool just ahead of me. No take. I stepped into the cobblestone-lined hole up to my knees and cast upstream again. As I carefully worked the fly toward me, I felt the line straighten in response to a strong tug. I set the hook firmly. A 19-inch rainbow skyrocketed three feet out of the water ten yards in front of me. Upon hitting the water, the frantic fish zigzagged across the pool five or six times until I managed to pull it within ten feet of where I was standing. As I reached for my net the fish lunged desperately straight upstream and danced on the surface several more times. The reel on my rod sang as the fish pulled gallantly. A few more zigzags and this time I managed to maneuver the fish within four feet of the outstretched net. The trout made a few more lurching plunges but he was tiring. A minute later I had the fish in my net and placed him into my fish bag attached to a suspender of my waders. The homemade fish bag was long enough that the rainbow could remain in the water as long as I stood in water at least a foot deep. "*It was going to be a good day*," I thought to myself.

"*No use trying this hole any more*," I thought, so I headed downstream. I carefully caught and stored two more handsome fish, a 20-incher and another 19-incher and I released several smaller trout. Only an hour had passed and already I had caught and landed half my limit and released several more that would be keepers in most fishing waters. It was time to light up a cigar.

Then I remembered that I hadn't caressed the fly that I was using with tobacco-scented fingers earlier. "*It'll be interesting to see if the cigar smell makes the fishing better or worse*," I speculated. I worked my way down to the "Elbow Hole." Walt had always referred this way to the hundred yard long hole with a sharp bend in it. I caught my 5-½ pounder from the Elbow Hole.

As I gingerly pushed my way through the watercress and moss into the deeper water, now encasing me up to my belly, I tossed my fly into the open water ahead of me. The fly had hardly sunk when I felt a light touch. I jerked and brought up a ten-inch rainbow. I unhooked the juvenile fish and cast again a few feet below where I had last released my Big Shrimp Special a half minute earlier. As the fly sank and began to drift slowly downstream, I felt another light take. I gave the fly a firm tug but nothing moved. Then my rod bent and my reel screamed as line rapidly peeled from the spool. I tightened the drag but the line continued to disappear. The fish was already into my backing. I started to move downstream. I hoped I could turn the fish before it reached the end of the pool and headed downstream through the fast running riffle. I raised my rod even higher and pulled with all the force I could use without breaking the six-pound tippet. Finally, no more line was going out and I began to rewind the reel. Slowly I gained a few feet here and a few feet there as the fish swam from one side of the pool to the other, dove and occasionally lumbered to the surface. I could see the fish's broad sloping back and I knew I had a big one on the line. Would it top my 5-½ pounder from several years earlier?

The battle seesawed another ten minutes. Each time I gained line, the big rainbow made another run and stripped off line. Each successive run was a little shorter than the previous. I clenched the cigar between my teeth and shook my head to spill the ashes because I had to use both hands to fight the big fish. Several more minutes passed and the moment to try to net the fish was approaching. On the first attempt, as the fish glimpsed the net extending toward him, he lunged desperately and deeply. The fish tried to entangle itself in the nearby watercress but I was able to keep the fish facing me so that it couldn't wrap itself in the vegetation. Two more attempts with the net to scoop up the big fish ended in similar circumstances. I uttered a brief prayer to ask if God really wanted me to have this big fish. On the fourth scoop, I managed to tip the heavy front end of the big fish into the net and then I quickly uplifted the net and the rest of the body fell into the mesh. "What a great fish!"

I uttered a quick thanksgiving prayer and carefully stepped to the shore. After removing the fly from the still-netted fish, I laid the net and the fish on the rocky shoreline and pulled out my portable scale with a retractable tape to measure the fish. It read, "Five pounds-eight ounces and 24 inches." After stuffing the huge fish into my fish bag, I sat on a big rock to finish my cigar. Eventually, I went back to fishing and smoked another cigar while I netted two more 18-inch, bright red and silver rainbows and released several smaller fish. Gradually I worked my way back to the entry point, all the time casting into pools that Walt had variously named: Spring Hole, Long Stretch, and Green Hole. It was time to eat lunch and have a cold soda that I had packed earlier this morning.

Just as I was climbing the trail heading out of the canyon, I heard a vehicle pull up. I couldn't see the vehicle over the cliff rim, so I kept tromping upward until I managed to glimpse a brown pickup truck with a picture of Idaho and official-looking lettering on the side. I wondered to myself, *"Did I keep too many big ones?"*

A cheery voice rang out as a dark-haired man wearing green sunglasses emerged from the truck and surveyed me and the heavy bag of fish I was carrying. "Rob's the name and I see you have good fish, a good cigar and it's a great day! It doesn't get any better than this, huh!"

"What a relief," I thought to myself as I realized that the friendly fellow only meant to share in my pleasure and joyful accomplishments. I continued to wonder if there was a newly imposed limit on the number of big fish one may keep from this stretch of river.

Over the next few minutes Rob and I exchanged enough information for me to find out that Rob was a licensed outfitter and guide from Sun Valley. Rob told me there was no size limit on the fish from this river. Nodding toward the Jeep with Iowa license plates, Rob wondered how I knew about this great river. I explained that my father-in-law had discovered it decades ago and had passed along the knowledge to me and now to my son, Jon. Only a few of Walt's friends knew about the river. Rob explained that he never brought his clients to this fabled fishing site because he didn't want others to know about it and to ruin it.

After eating lunch and watching Rob get ready to fish, I offered Rob my last remaining cigar. I explained that I had already smoked two cigars. Rob declined, saying, "It's a three cigar day, don't you think?" I nodded in agreement.

The Sins of Omission Are Worse Than the Sins of Commission

As a youngster I spent a good deal of time in the confessional for my various transgressions. My elementary school report cards were filled with misbehavior checkmarks: "lacks self control, doesn't use full abilities, challenges authorities." Raised a serious Catholic, I could be readily motivated by guilt.

While I waited in line at the confessional, I often debated which sins were venial versus mortal sins. I rehearsed what I would tell the priest in the confessional, "Father, forgive me for I have sinned. This past week I lied four times, took money from my mother's purse without asking once, hit my brothers six times, and had impure thoughts about Dolly three times, one of which was a mortal sin."

Dolly, who was two years older and hot, was ahead of me in line to enter the confessional. When Dolly entered the confessional and everybody in the waiting line moved ahead one step, I tuned up my auditory reception so that I could possibly overhear the sins that Dolly whispered to the priest. I figured that if I inadvertently overheard Dolly say that she and Sam had a romp in the hay last night, it was only a venial sin. If I tried really hard to listen to the juicy parts and started to fantasize that it was me she was doodling, I believed it was a mortal sin. I had been taught that I would be doomed to hell if I died before confessing a mortal sin. But my fate was protected at my young age because, after all, I was next in line to confess my sins. The slate of my conscience would soon be wiped clean.

These acts of commission were easy to identify and made decisions about what constituted moral turpitude simple. As I was becoming a serious outdoorsman in my adolescent years, my hunting and fishing activities were largely governed by my conscience. I reckoned that if I kept four bass over the minimum size and the state law said the limit was three, clearly I had committed a sin. I visualized conservation officers swooping down on me from all sides with guns aimed at my chest. It felt wrong!

By my early twenties I was beginning to wonder what it was about fishing that made it such a wonderful pursuit. Fishing activities were symbolic of my overall approach to life. As I matured, it became less important how many fish I caught and more important how I caught them. Were they caught on flies that I tied versus those that were given to me? Even more significant, were they purchased from a company that employed child laborers in China who were paid three cents for each correctly tied imitation of a caddis emerger? Not that these internal debates were unhappy times. Just the opposite—thinking about the ramifications of the fish I caught and how they contributed or detracted from the overall good of the whole could be, and usually was, a pleasurable internal discussion. It was satisfying to know that catching only enough to eat, or cleaning and offering fish to the owners of the ponds that I was allowed to explore was a "giving" behavior. It was important how I fished and hunted, with whom I enjoyed these experiences, and how these personally restorative activities could enable me to work better and keep my overall motives purer.

I began to realize that I am but a small component of a much larger system and my role is to contribute to the good of the whole. If everyone contributes to the good of the whole, all benefit. If nearly everyone detracts from the good of the whole, the system will collapse. Life is a collection of sustaining and detracting acts. Behavior tilts on a fulcrum: it is either more "contributive" than "selfish or subtractive" and vice versa. That we all have different views about what is good for the whole is something to ponder. These emerging views made my decisions about right and wrong more complicated.

Commission and omission! The sins of commission are usually simple, straightforward, and dumb. For example, it's not very difficult to figure out that most thefts, assaults, and adulterous relationships are selfish acts of commission because they harm other people. But it's a lot harder to determine that the modern-day practices of executives receiving huge bonuses that are many times greater than the average worker salary in the company, or failing to correct negative campaign distortions of election opponents' records, or undertaking plastic surgery just to earn admiring comments are sins of omission. For nearly all who undertake these acts, their aim is mainly personal gain. Our motives are critically important. Are our motives for personal gain or for the common well-being?

Over the years I have come to realize that more is expected of those who have more. Money, power, attractive physical looks—they are all gifts that if rightly applied, can contribute enormously to the good of the whole, or they can be used downright selfishly. My younger brother,

Larry, was a good example for me of someone who had a pure and rich life because he was almost entirely unselfish. He was born eight years after me with reduced capacities from Down syndrome and a bowel imperfection that required surgery when he was three days old. Larry could "make do" with simple things. Though he often experienced considerable pain, he did not complain. He found the perfect words within his limited vocabulary to express jubilation when he felt happy and would blurt out in a husky voice, "Excellent joy!" Even on his death bed, in May 2008, Larry accepted what was given him and used his suffering to make himself a better person, and those who witnessed Larry struggle for breath.

Larry was hospitalized the last few days of his life. Gradually he became weaker and his breaths became more uneven during the last few hours. My brother and his wife, Maria, Marilyn and I, and a nephew were with him most of his last hours. We talked to Larry and among ourselves of his imminent death and said our good-byes. Larry could hear us, for he responded with gestures. Larry didn't become scared even though he seemed to know what was happening. It seemed as if his accepting manner was conveying, "I can do this, you don't have to be afraid or worried about me." We were with him continuously from about noon the day before, through the afternoon and night. His breathing became somewhat more stable around 5:30 AM so most of us except Maria went home to get ready for the day. Larry died just be few minutes after we left his hospital room.

Larry was also a factor in my decision, early on, to become a psychologist. Watching him while I was still in high school, I knew I wanted to help people. I wished I could get it right as easily as it seemed to come to Larry.

I began to realize that I am my own worst enemy in terms of doing the right thing and for the right reasons. While recognizing and managing my motives has influenced every aspect of my life, I want to focus a bit on fishing. With fishing, it is easy to know when I have committed a selfish deed, such as catching fish out of season. But even in this realm it's the omissions that really define us.

While most omissions are difficult to perceive, some are easy to detect. A common omission that I have repeated many times is forgetting an essential fishing item. Usually I didn't discover this until I had driven 40 miles over dusty winding roads and pulled alongside the shore of a promising farm pond, that looked to be just brimming with hungry bass and crappies. I had my waders, float tube, fly rod and reel, fish net, hat, even the insect repellent, and cigars. But I wasn't going anywhere on the pond without my fins, which were still laying on the top step of the basement stairs.

Forgetting lunch or drinks is another common mistake. Not having a beer or soda or something to eat has to be balanced against continuing to fish. Not having something to eat or drink doesn't necessarily interfere with fishing except during the occasions when Ken was along. Ken, my graduate school roommate before Marilyn and I married, quite frequently injected unpredictable subtleties into fishing. Usually he didn't intend these complications—they just happened.

A few years back when we were fishing Montana's Madison River in the early '90s, Ken chose to carry the sandwiches and I agreed to carry several beers in my insulated backpack. We separated at an agreed upon "put-in" point at 9:30 AM and planned to meet back there at noon. Ken's vehicle was parked nearby. Ken headed upriver and I headed downstream. By late morning the sun blazed hotly. Every so often I pulled out a cold beer to celebrate catching a nice fish and to cool down. By noon I was back at the put-in point but Ken was nowhere nearby. I whiled away additional time casting into close-by pools and landed a couple more fish but they weren't as big as the five I had in my creel so I returned them to the river. One more fish and my limit would be complete.

By 12:30 PM it was clear that Ken was somewhere else, for he didn't answer any of my calls, so I headed upriver. Besides, I was hungry even though I had consumed several beers. And Ken would be thirsty, I thought.

Almost an hour later and a couple thousand yards upstream, I spied Ken fishing a quiet stretch of water behind a beaver dam between two islands. As I trudged closer, I yelled to Ken, "How are you doing?"

"Rosmann, what took you so damn long," Ken bellowed with obvious irritation.

"I waited half an hour for you where we put in and finally decided to come looking for you," I responded.

"The hell you did," Ken loudly stated. "We were supposed to meet right over there," gesturing with his head toward a distant shore.

"But your car was down where I waited," I bravely retorted, "not over there," while pointing my rod toward the shore.

Realizing that he had omitted keeping track of his bearings, Ken weakly defended himself, "Oh well! I've been having a lot of fun but I'm thirsty as hell."

I traded Ken a beer for a sandwich and asked, "Did you get any nice ones?"

"Oh yeah, check these out," Ken advertised as he opened the lid of his creel to expose five, twelve to fourteen-inch brown and rainbow trout. "How did you do?"

"I threw most of mine back," I responded as I reached into my creel and pulled out an 18-inch brown trout and brushed off the wet moss I had stuffed among the fish to keep them cool. Slowly I laid it on the dry gray rocks of the streambed below the beaver dam and carefully laid four larger trout beside it.

Ken gasped, "What are you using?"

"Oh, I've had my best luck with a brown stonefly that I tied a few years ago. I'll give it to you."

"You don't have to do that," Ken volunteered quietly but in a tone of voice that I knew to really mean, "I would be very glad if you would give me your fly."

While Ken ate I replaced the fly on Ken's rod with my own creation and gave Ken his rod back. In between sips of beer, Ken made a few casts into the beaver pond just above where we were standing. No luck.

I had finished my sandwich but Ken was still washing down bread and bologna with his third beer. "You're not letting the fly sink right," I suggested. "Let me show you."

Ken handed his rod to me. I stepped into the beaver pond a foot or so and with a delicate roll cast dropped the fly upstream 20 feet and let it drift toward the sticks that the beavers had stacked into a sturdy and nearly leak proof dam. As the submerged fly drifted within a couple feet of the dam, a bright yellow and brown spotted trout emerged from its lair nestled among sticks and mud deep in the beaver hole and seized the fly firmly. I set the hook hard. The trout bolted upstream in a brisk run that peeled line from the reel while the clicker sang loudly. "Here, take your rod," I offered.

"Go to hell," Ken loudly pronounced and stomped away. Ten minutes later I landed a 23-inch brown trout, the largest of the day. I offered it to Ken, but Ken was royally ticked. I realized that I had omitted consideration for Ken's pride. But then, Ken had omitted keeping track of the time and place. Now Ken was particularly sore at me and I felt badly. Dang those sins of omission!

A day earlier, on the same trip, Ken and I had arrived at a ranch about two miles from the Madison River where we supposedly had permission to camp overnight. A thunderstorm cut loose. Ken mentioned that he had helped take care of the rancher's son in March this spring at the hospital in Denver where Ken worked. The rancher's son had severe asthma. Ken said the rancher offered to let Ken camp on the ranch whenever he wanted to fish the Madison River.

Small hailstones and big drops of rain splashed on the grass for nearly ten minutes. When the storm cleared, Ken and I strolled up to the front door of the ranch house and gently knocked. No response.

As we looked around and could see that there was no vehicle parked in the driveway and that probably no one was home, Ken suggested that perhaps we could pitch our tent in a dry spot somewhere. He motioned toward a barn with an open front about 150 yards away. We hopped into Ken's Toyota and drove to the barn. As we were setting up the tent, a truck drove up and a stern-looking burly man wearing a plaid shirt and brown cowboy hat demanded, "What the hell do you think you are doing?"

"I'm Dr. Benson," Ken replied.

No response. Only silence for several seconds.

"Remember, I helped take care of your son, Eric, when he was in the hospital in Denver this spring."

Still silence. "You said I should come see you whenever I wanted to fish the Madison," Ken stammered.

Gradually the angry looking man's face relaxed. "Oh yeah, you're the guy who did therapy with my son. I guess I did say something like that, didn't I?"

Over the next ten minutes of small talk, it gradually dawned on Ken that he should have called the rancher to let him know that he was coming to Montana and to reaffirm the offer to let him camp there. The tension dissipated after Ken apologized. The next evening the rancher and his wife invited us over for grilled elk steaks that I must say were the finest elk steaks I have ever eaten.

It usually is easier to spot someone else's omissions than our own. And so it seemed to me that omissions were a common part of Ken's life. Usually they were the source of much good natured merriment. There was the time Ken forgot to gas up his vehicle that he, his wife Carla, Marilyn, and I took to Evergreen, Colorado, to visit a restaurant there. By the time our dinner was done, all the service stations in Evergreen were closed. Somehow we managed to coast all the way down the eastern slope of the Rocky Mountains on Interstate Highway 70 until we rolled into a busy service station in west Denver without having to crank over the engine once.

And then there was the time Ken forgot to add washer fluid to the reservoir in his Volkswagen. Ken and I were returning from a rock-hunting and fishing trip in western Utah. Ken had driven through a deep mud hole and splashed thick brown mud over his entire windshield. Even with the wipers swishing back and forth furiously, all they did was move the mud from one spot to another.

"Stay here," Ken said as he piled out of the Volkswagen squareback he was driving. He crawled on top of the hood and clambered onto the roof. The next thing I saw was a steady stream of yellow-green fluid dribbling onto the windshield. Ken yelled for me to turn on the wipers, which I did. Mud flew in all directions and the windshield became clean. The ammonia in Ken's pee was good glass cleaner. What do you know—here was another omission that turned out delightfully okay! Some events didn't turn out as well.

Several years later Ken invited me to go fishing on the headwaters of the Rio Grande River on the western slope of Colorado. It was a hot August afternoon. Ken had packed his Toyota with fishing gear and a few camping items. I stowed my fishing and camping gear inside the Toyota and we set off from Denver. After just a few miles we noticed wisps of steam coming out the back of the Toyota's hood. Ken pulled onto the highway shoulder. After raising the hood, he could see that steam was gushing out of the radiator cap. Grabbing a rag, Ken leaned over the radiator to push down on the cap so as to release the pressure inside.

"Wait," I yelled, but it was too late.

The pressure blew the radiator cap out of Ken's hand. Steam and scalding hot antifreeze blasted out of the radiator directly onto Ken's face.

"Damn! That's hot," Ken gasped.

I quickly grabbed a thermos filled with cold water intended for the camping trip and splashed the cold fluid onto Ken's hands and face. "Make sure you get it out of your eyes."

As Ken washed the coolant fluid from his red face, I checked to see if he was blistering. Fortunately, Ken's burns weren't too bad. He explained that he forgot to add water and antifreeze yesterday after replacing a cracked radiator hose on his Toyota jeep.

I poured the remaining water into the radiator and the gurgling inside stopped. It took the entire second container of camping water to fill the radiator.

The omissions on this fishing trip were not at an end yet. Five hours later as Ken and I drove to a U.S. Forest Service Campground along the upper stretches of the Rio Grande River, Ken's Toyota began to run unevenly. Ken pulled into the first available campsite just as the vehicle sputtered to a stop. As Ken tried to restart the engine, it was obvious that the battery was dead. After raising the hood and inspecting the wiring, I discovered that the wire from the battery to the alternator was not connected.

Ken hooked up the battery cable properly. "I forgot to do this when I was working on my vehicle yesterday," Ken mentioned lamely.

Pretty upset with the way things were going, I bit my tongue and suggested, "Let's just camp here." Then, I added, "Don't try to start your vehicle for awhile. Sometimes the battery will rejuvenate a bit if you just let it set."

The omissions turned into commissions. Ten minutes later while I was setting up the tent, Ken jumped in his vehicle and turned the switch to start the engine. The engine turned over weakly twice, followed by the ominously limp clicks of a dreaded dead battery. Scouting around, I spotted no other persons from whom to beg a jumpstart. I wanted to cuss but knew it would only make matters worse. Then an idea occurred to me.

"Benson, get out of your vehicle and help me push your car." I put the gearshift into neutral. With all our strength we pushed the vehicle backwards up a small incline. I shoved a rock under the closest tire to keep the vehicle from rolling down the slight grade. Then I put the gearshift into reverse, set the emergency brake and took the keys!

It was obvious to Ken that I was thoroughly ripped. It would be better to leave me alone for awhile, Ken probably thought. He was accustomed to my temper and strong will.

As I finished setting up the tent, Ken made supper. Few words were spoken over supper. Night was approaching fast. I suggested that Ken check out the stream before it was too dark to see. As Ken meandered toward the creek a couple hundred yards away from the campsite, I pulled the Toyota keys from my pocket and kicked the stone out from under the tire. I jumped behind the steering wheel, turned the key to "on" and put the gearshift into 2nd gear. I held the clutch down so the vehicle could coast. I was really glad this was a stick shift transmission because what I was about to try would not work with an automatic transmission.

I released the emergency brake, and then with my left foot shoved the vehicle forward while holding the clutch pedal down with my right foot. As the Toyota began to roll down the slight incline, I steered it onto the asphalt road, which also had a slight downward grade. After coasting a couple hundred feet and picking up speed, I let the clutch back, turned the key to "on" and pressed on the gas pedal. The clutch grabbed, the engine turned over several times and sputtered to life.

"Whew!" I pulled the vehicle onto the shoulder of the road and sat there, with the engine racing for several minutes to pump life into the battery.

I drove the Toyota back to the campsite and parked it just as Ken was getting back from the creek. "How did you get this thing started?" Ken inquired. I explained how I learned to start tractors this way as a kid on the farm. Things were getting better!

My father-in-law gave me useful cues about how to sort through my motives regarding fishing and life in general. Just as for Walt, fishing became an opportunity to improve the common good as well as to help me. One April, about 20 years ago, after three weeks of 16-18 hour long days without a break, I came to realize that Jon and I had not gone fishing for awhile, even though it was spring. I hadn't even purchased a fishing license yet this year. It was Sunday. No vendors from whom I could purchase a license were open for business today.

"Should we go fishing anyhow?" I mentioned to Jon over Sunday brunch. "Yeah," Jon replied excitedly. Jon didn't need a license—he was still younger than the minimum age of 12 years after which a license was required.

Still pondering whether or not to go fishing without a license the entire way to Keene's pond some twelve miles distant, I realized that I had been neglecting my family and myself lately. While I was working excessively long hours I had omitted adequate recreation. I knew I was grumpy, tired, and needed a break. *"So what if I get caught,"* I thought, as I considered the ramifications of not owning a fishing license. That a game warden had never checked me for a license during the ten years that we had lived in Iowa helped me feel braver. I needed to go fishing for the sake of everyone in the family. Marilyn gave her blessing.

I parked the pickup truck, loaded with our canoe, next to Keene's pond and we readied our fishing gear. When the canoe was loaded, I shoved off the shore. Jon always took the front of the canoe while I took the rear. We had a well coordinated method of transportation and we knew each other's habits. We never cast simultaneously so as to avoid tangling our fly lines. We took turns holding the canoe in a favorable position so that the other person could cast into particularly promising openings between weed beds. Usually that meant the person who was temporarily in charge of paddling could fish, but he always relegated his fishing to maintaining the canoe in a favorable position. When it was really windy, holding the canoe in a favorable position was a full-time job and that person couldn't fish for the time being until it was his turn to cast for awhile. It was common for Jon and me to fill a five-gallon bucket nearly full of bass, crappie, and bluegills during a couple hours of dedicated fishing and careful canoeing.

Hardly had the canoe left the shore when Jon caught his first hefty bluegill and heaved it into the fish pail. A minute later I had captured another frisky bluegill and also tossed it into the five-gallon bucket.

Just then we heard a vehicle on the gravel road above the dam on Keene's pond. It came to a halt when the driver spied Jon and me in the canoe.

Jon recognized Officer Dick Johnson, the District Fish and Game Conservation Officer who had taught his gun safety class last fall. Officer Johnson gestured with his hand to bring the canoe ashore.

"May I see fishing licenses, please?" Officer Johnson asked. Jon waited for me to answer.

After a short pause, I responded, "Jon's not 12 yet and I don't have a license."

"I'm going to have to write you up," the conservation officer pronounced calmly.

"That's okay," I answered politely. "I made the decision that we were going to fish even though I didn't have a license. There isn't any place I could buy one today. I've been working too hard lately and we needed to go fishing."

"I know what you mean," Officer Johnson replied. He took my name, address and telephone number from my driver's license, wrote out the ticket and handed it to me.

"Thank you." I said and then asked, "Would it be alright if we continued fishing here today?"

Officer Johnson paused for a moment and then said, "Let's just say I won't be coming back this way anymore today."

Jon and I both said, "Thanks." We watched Officer Johnson crawl back into his vehicle and drive off. Then we paddled the canoe away from the shore and returned to fishing. I realized that Officer Johnson also appreciated that committing an illegal act in this case was better than omitting what was more important.

The Red-Butted Green Nymph

Gladly, I steered my white Cherokee Sport Jeep onto the gravel driveway of "*The Pines*," a congregation of 15 clapboard cabins strewn among massive ponderosas, oak, ash, and assorted shrubs gracing the slopes overlooking Pine Creek Canyon. I had reached *The Pines* by mid-afternoon on this sunny first day of June. It was my favorite place to hole up in Nebraska and, arguably, one of the prettiest spots in the state. I had driven hard for seven hours to reach Pine Creek from Cheyenne, Wyoming, where I had given a conference presentation the day before.

As I pulled to a stop in front of the proprietor's cabin, Chad stepped onto the porch to greet me. Chad and I had come to know and like one another over the past several years. In his mid-twenties, Chad was an aspiring fly fisherman and fly tier. I was eager to unpack, to purchase my fishing license from Chad and begin fishing Pine Creek. That the trout in this stream were often difficult to entice to a fly made fishing Pine

Creek all the more fun. There were many hidden overhangs along the stream banks to explore, dark holes where the water would flow over one's waders if you stepped in the wrong place, and promising ripples around the multitude of creek boulders, some as big as hay bales. There was an abundance of wildlife. Noisy turkeys just outside the cabin often awakened me at daybreak. Deer and foxes skulked in wooded areas and the grassy fields along the creek. Once, I had spotted the tracks of a mountain lion in the soft sand next to the creek.

When Chad told me that a well-to-do Easterner had purchased this stretch of the valley and would not allow fishermen access to Pine Creek, I was disappointed. My disappointment worsened when he told me my cabin wasn't ready for occupancy. He had been kept overly busy attending to the needs of a dozen members of the Nebraska Chapter of Trout Unlimited. With triumph in his voice Chad announced they were holding their annual meeting at *The Pines*. Chad's enterprises needed an economic boost and the Nebraska Trout Unlimited Chapter could provide just the sort of boost he needed. Judging from the make and size of the vehicles parked by their cabins, the NTU members were well heeled. Plus, there was an off-chance that they might buy some of Chad's fly creations and he could probably learn some fly-tying tips from them.

I hoped that *The Pines* would not change much. I liked the accommodations the way they were—wavy cabin floors, refrigerator doors that were held shut by screen door springs, bed frames made of pipes and constructed by plumbers, and best of all, the place was relatively undiscovered except by the eastern tycoon who now restricted access to Pine Creek. While I was pondering this, I noticed a man who resembled Andy Griffith in his fifties stroll up the lane to consult Chad. He prominently displayed the Nebraska Trout Unlimited epaulet on his fishing vest. Sensing the opportunity of the moment, I proclaimed, "I'm a member of Trout Unlimited too."

The silver haired angler approached my vehicle. "I didn't know that you had trout in Iowa," he pronounced, motioning toward the Iowa license plates on my Jeep.

"We have some respectable trout in northeastern Iowa but Pine Creek is actually closer to where I live."

The Nebraska Trout Unlimited fellow announced that his name was Jim and he was president of the state organization. Jim said he and his fellow members had visited fishing ponds at Keller State Park, about ten miles away, earlier in the afternoon and had good fishing. Chad suggested that I ought to enjoy myself fishing while he readied my cabin. The NTU president volunteered directions to the park and invited Chad and me to the annual meeting of the NTU Chapter that evening around a campfire outside his cabin.

Three hours later I had caught and released a dozen 10-12 inch trout, kept a few for supper and breakfast and returned to find my cabin ready. Following supper I selected two flies from my assortment to present to Jim at the NTU Chapter meeting as a gesture of appreciation for his advice about Keller Lake and for the invitation to the annual meeting. One fly was a black creation that I had invented and used many times to catch bluegill, bass, and crappie. I hoped the NTU president might be interested in catching other species than trout. The other fly was one that I had learned to tie from my father-in-law, Walt. It was a nymph pattern with a green mohair body, brown hackle, and a touch of red wool on the rear end. When I presented the two flies to Jim at the campfire, he graciously thanked me and asked what the flies were called. I explained that I called my invention, "the little black fly" because that's what my kids called it and the other fly had no name.

"Oh," the NTU president exclaimed, "and this other fly looks like a red-butted green nymph. Is it any good?" he asked skeptically.

"Yeah, it works pretty well on some lakes in Idaho but I haven't tried it here. My father-in-law showed me how to tie it."

The annual meeting agenda consisted of fishing stories that became increasingly more colorful, funny, and outrageously inaccurate as the night progressed and as the beer and scotch flowed freely. Chad was enthralled and I mostly listened. Jim told how he caught and released Apache cutthroat trout in New Mexico last month. Ruthie reported catching salmon off the coast of Denmark last fall, and Darrell described his trout fishing successes in Argentina and New Zealand in January and February. The NTU members compared rods, which seemed to become more expensive and rare as the night passed. Jim owned an extensive collection of *Orvis* fly rods. Ruthie used only hand constructed *Thomas & Thomas* rods and *Hardy* reels. Charlie had a one-of-a-kind hand split bamboo rod that had been passed down from his ancestors and which was his favorite. The NTU members weren't impressed when I mentioned that my favorite combination was a Medalist reel that my father-in-law gave me as a Christmas gift nearly 30 years ago and a nine-foot graphite rod my son gave me two years ago for my birthday. Chad's fishing equipment was even less pretentious—he owned only two fiberglass rods and a few flies that he had purchased or constructed himself.

Perhaps out of kindness for Chad and me, the NTU folks invited us to join the Chapter for an outing at a member's nearby ranch the next morning. "Just be ready to leave before 7 o'clock and you can follow us," Jim said. We gratefully accepted the invitations as we said, "Good night," and stumbled toward our cabins.

I was in my Jeep by 6:45 the next morning, with all my fishing gear and bags packed. I intended to drive home to Iowa after our fishing excursion. My plain white vehicle looked awfully humble among the shiny black Expeditions, chromed Yukons, and fancy polished Land Rovers. Chad opted to ride with the NTU president and seemed mighty pleased to be, "up there with the dignitaries."

After a 30 minute ride down several Nebraska highways and a dusty gravel road, we pulled into the ranch where the group was invited to fish a private pond stocked with Kokanee trout. The ranch owner explained that he wanted us to catch and keep as many fish as we could because there were too many in the pond and he wanted to thin out the population.

With high expectations, the troop followed the ranch owner through his fields to a spring-fed, man-made lake about three acres in size toward the back end of a tree lined pasture nearly a mile from the homestead.

It took me only a couple of minutes to pull on my waders and string up my rod and reel. I decided to use my favorite slow sinking line because, as Walt once advised, 90 percent of what trout consume is under the surface. I was already casting into the far end of the lake by the time the next fisher stepped up to the water's edge next to where the vehicles were parked.

On my second cast, I felt a sturdy yank and soon my rod was bent nearly double and the line taut as a good-sized kokanee gamely swam back and forth ahead of me. I played the fish carefully. By the time I netted the chunky fish, several minutes later, most of the NTU members were strung out along the shore directly across the pond from me. Some had watched me land the fish and exchanged comments. "There must be some nice fish in here. What fly do you think I should use?"

After unhooking and bagging the stocky 19-inch kokanee, I lifted my rod and line vigorously and in a single, fluid, backward swoop and forward motion threw the fly 60 feet toward the center of the lake. Again I noticed several of the NTU members eyeing me.

The NTU members cheered each other on, "How about a royal coachman on a floating line today," one NTU member urged his neighbor. Another nimrod whipped his pole back and forth in seven-eight furious false casts and finally dropped the fly about 25 feet in front of him. A small chorus of, "Nice cast!" emerged from admiring bystanders. The NTU Chapter was about to give the neophytes, Chad and me, a lesson in fly fishing.

As I quietly fished by myself at the south end of the lake, shaded by several large trees in the background, the NTU members gradually fanned out along the north, east, and west shores. Several casts later, I had another solid hit. The fish pulled so hard that I had to release all the fly line I had coiled in my left palm. I recovered line gradually for several minutes while spectators around the lake surreptitiously glanced in my direction. After several minutes I netted a broad shouldered 20-inch kokanee. Quickly I placed the beautiful dark silver fish with purple, red, and green coloration on its sides into my fish bag that was suspended in the water next to my left leg. Again, with a single backward motion and forward thrust of my right arm, my fly sailed gently onto the surface of the pond 60 feet ahead of me.

Someone had hooked a fish across the lake and was gamely playing the kokanee on his reel. "Keep your rod tip up," a high-pitched male voice admonished. "Let him take some line if he wants to," another man volunteered. The thrashing fish was giving Darrell a tussle. Darrell pulled the fish toward him and lifted a ten inch fish out of the water. Just as Darrell was sweeping his net toward the fish, the trout wiggled free and flipped into the pond. "Darn it all," Darrell loudly proclaimed. Others joined in with, "Too bad, Darrell," and, "Better luck next time."

Over the next half hour, I caught and released four more fish; the smallest was 15 inches long and weighed over a pound. I shifted a few feet to the right several times to try out fresh water. Fishermen and fisher-women gingerly spread closer to me along the shoreline.

Periodically someone shouted, "I got one," only to be followed by con-solations when the fish unhooked itself a few minutes later. Someone uttered, "Good fish there Ruthie," followed by, "Oh fiddlesticks!" when a nice kokanee broke Ruthie's tippet.

From time to time, however, an NTU member managed to land a fish and many comrades sauntered over to congratulate the proud victor.

As the morning moved along, I landed and bagged two 21-inch trout and released several smaller fish. Most of the fellow fishers crept within a hundred feet on both sides of me. Chad planted himself only 20 feet off my left side and a tall slender erudite man with a British accent and a plaid tweed cap experimented 30 feet from my right side. I noticed that Chad regularly cast into the spots that I had vacated and that several fish had nipped at Chad's fly but none had firmly grasped the bait. As Chad and I made eye contact he slid over to me and asked, quietly but loudly enough for several nearby persons to hear. "How many fish do you have?"

"I think I kept four."

"Four!" Chad pronounced loudly enough that everyone could hear, "That's more than the entire Nebraska Trout Unlimited Chapter!"

I quickly reeled in my fly, snipped it off the tippet and held it out to Chad. "Here, try this."

Briskly I hiked toward my vehicle, making sure that I stayed at least 40 feet behind the fishers who were vigorously flicking their lines back and forth in violent false casts. As I passed behind Ruthie, she stopped casting and shyly inquired, "What were you using? I saw you catch some nice fish out there."

Reaching into one of the zippered pockets of my vest I pulled out a cracked, round, clear plastic box that once contained a spool of electrician's tape and now was packed with hand-tied flies. I handed Ruthie the same kind of fly that I had given Chad. "Good luck!" I hailed.

When I reached my Jeep, I quickly gutted the four fish I had kept and placed them in a plastic bag in a cooler. Just as I pulled off my waders and slipped on hiking shoes, I saw the NTU president stalk toward me, saying, "You were having a lot of fun out there, huh? What were you using?"

"Oh, a little something I tied up myself," I replied. "What were you using?"

"I tried just about everything, but I didn't catch anything yet," Jim responded. I reached into the fishing vest I was still wearing as I crawled behind the steering wheel of my jeep. I handed a fly to Jim and turned the key to start the engine. "Well, I'll be danged," the NTU president said, and then added, "the red-butted green nymph!"

I stepped on the gas and watched Jim through the rear view mirror. I could see that Jim's face was nearly as red as the rear end of the fly I had just given him. Over the next few weeks I sent several emails to Jim thanking him for inviting me to visit the rancher's private lake and to wish good tidings to the Nebraska Chapter of Trout Unlimited, but he never answered back.

Where There Are Bonds

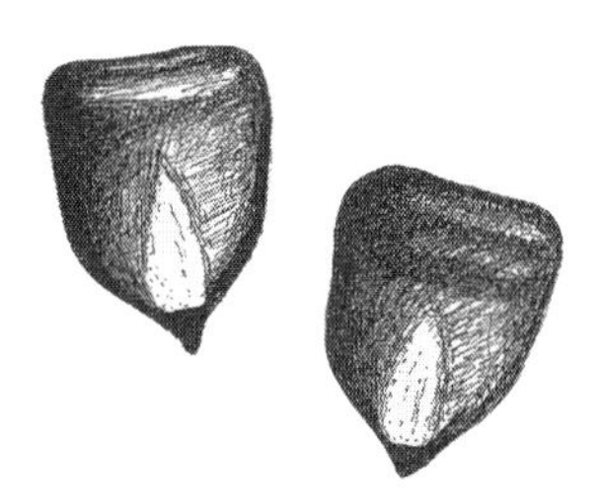

One night, several years ago, something happened on our farm that affected me profoundly and it seems likely to influence me into the future. The event also had a powerful impact on Jon, then 15. It was a chilly drizzly day in late March, the kind that one just endures. This was the sort of Saturday when you want to sit in the house close to the fireplace or television, watching basketball or wrestling tournaments.

After a busy week of work at the mental health center, I was ready to collapse into my recliner in front of the television. Larry, our hired hand, had put in his obligatory 32 hours already during the week, so earlier today Jon and I carried out the unfinished but necessary chores.

We shoveled manure out of the barns and replaced soiled bedding in the calving pens with fresh straw. We ground two wagonloads of ear corn, one for the "east place" and another for feeding cattle at home. We fed the ground corn to boost the nutritional level of our remaining stored cattle feed. Most of our good alfalfa/orchard grass hay was gone. We were feeding bleached hay that had been rained on before it was cured last summer and big bales of conrstalks, which were low in energy and nutrients, along with the necessary corn supplement, to carry the cattle herd until the emerging grass grew tall enough to graze.

It felt good to work hard physically. My dual callings to be a psychologist and a farmer blended together and enriched each vocation. Farming provided physical outlets, communion with nature, and the nurturance to refill my social emotional fuel tank, while my psychology work helped me to approach agriculture scientifically, to think about farming in interdisciplinary ways that I would not otherwise have engendered and to lay the essential groundwork for a new field: agricultural behavioral health. The synergy of two primary pursuits could also be wearing, but the sheer joy of the synergism was energizing.

Every two hours while carrying out the farm chores, I kept an eye on the pregnant cows and heifers. I was into my usual calving season practice of getting up every couple hours at night to check the cattle that were ready to deliver their offspring. My visits to the cattle yards and sheds were timed with REM (rapid eye movement) sleep periods that usually were about two hours long. Spring calving season was well underway and we already had 25 of the expected 65 newcomers on the ground. We also had a smaller fall calving herd at our other farm. Most of the problems that seem to occur were out of the way early on, with the first few new calves. One was born dead and another was too big for his heifer/mother to deliver successfully, even though I helped her. He was a swollen-headed calf that didn't know how to suck his mother's milk. I drenched colostrum that we kept in our basement freezer, and had warmed to body temperature, through a tube into his stomach, but a day later he still laid next to his mother, floppy and with his eyes rolled up into his forehead. There wasn't much one could do to save him. Overall, the calving season was going pretty well. Taking a set of twins into account, we had a 96% success rate.

When we had completed all the required chores, Jon and I retired to the house at 4:30 PM I had not detected any potential deliveries for the next couple hours. I caught a few winks in front of the television. Jon went about his typical school-related activities and family pursuits, which usually included tormenting Shelby, his three year older sister. Marilyn played music loudly in the kitchen while preparing the evening meal.

Just before supper I donned my boots and rain poncho and tromped through half-foot deep mud to check the cattle lounging area for new developments and to see if the already delivered calves were snug in the open-fronted calf sheds with their mothers bedded down nearby on corn stalks and straw. An end-of-day, gray, foggy March gloom had settled over our farmstead. Only the bright lights of the house and the sounds of Marilyn's music broke the monotony as I scanned the horizon around the farm.

A big tan and white spotted first-calf heifer, ear tag A48, wasn't within eyesight. The progeny of one of my top lines of registered Simmental cattle, she was probably in the far corner of the calving pasture, just over the hilltop. I wanted to make sure she would be okay if she gave birth. Her mother and two full sisters comprised a family within the herd that had produced the top steer during the past two years of the Tri-County Beef Futurity.

The Tri-County Beef Futurity is a contest in which cattle producers enter three-five head of steers, all of which are fed a similar ration. When they are ready for market they are weighed and slaughtered at a nearby meat processing plant. Measurements of the quantity and quality of meat are taken on the hanging carcasses. The Extension cattle specialist for Southwest Iowa and a number of purebred and commercial cattle breeders organized the futurity to compare various breeds and genetic lines to develop recommendations for producers' breeding programs. The futurity winners were the top pen of three or more head and the top individual on two indices: retail value per day of age and retail value per day on feed. For the past four years in a row, our farm, Rosa Blanda Farms Ltd., had produced the top pen and the top individual from among 150-180 head of cattle. I usually sold my best bulls and some heifers to other producers, so I entered animals that had become steers into the futurity to feed out. I didn't want to retain them as bulls because they were not my best cattle. This particular line of cattle exhibited superior growth and meat producing capacity but they tended to have calves that were too big. I was searching for genetics that could reduce their birth weight while hopefully retaining their growth capacity.

Contemplating these prospects, I jumped onto the four-wheeler I had parked next to the field and scooted around the calving pasture. Just as I thought, A48 was lying on her side in the southeast corner, with a water bag hanging out her vagina. Raindrops clung to the fenders of the Honda ATV and droplets saturated the ground and vegetation on which the heifer was lying. I wasn't sure if I wanted to allow her to proceed with birthing here or to drive her 300 yards to the calving barn, where there was fresh straw and no rain! Sometimes it's better to allow nature to take its course and to sort out the animals lacking survival capacity, even if this family had superior-growth animals. If they can't have calves easily, this line of cattle would have to be terminated. I didn't want to produce breeding stock that could cause problems for other producers. On the other hand, if the problem wasn't the heifer's fault, she deserved the opportunity to see what she could do on her own—maybe she would produce the right genetic combination.

I turned off the four wheeler engine and its headlights and directed a spotlight toward, but not directly on, the heifer. I watched her silently. Her contractions were regular and strong. The calf's two front feet were protruding out her behind. They seemed to be covered with a membrane that looked like placenta. "*She should be able to deliver in ten-fifteen minutes,*" I surmised. "*I'll observe her for a while.*"

Five minutes went by. The rain-enveloped evening was comfortable though cool. I could still hear distant refrains of music from our house several hundred yards away. There was no wind, only rain droplets coating my poncho and penetrating the surrounding environment. A few minutes later it looked like the calf's nose was pushing out of A48's rear end. The membrane that stretched over the protruding front feet also stretched over the emerging nose. Each time the heifer had a contraction, the nose and feet bulged out and then retracted. I watched another five minutes, but no progress was occurring. It was time to aid the heifer and her calf.

As I walked toward the heifer, she raised her head to survey me, loudly moaned, and roused herself to stand. The legs and head of her emerging calf retracted. "I'm going to take you to the barn where I can help you if I need to," I announced hopefully. The spotted heifer looked at me without expression. I hopped onto the ATV, started it, and rolled toward the animal. She moved ahead of my vehicle and steadily strolled toward the cattle yards and barns.

As we approached the gate to the calving barn, I parked the four-wheeler far enough behind the heifer so she wouldn't get nervous and I hastily circled ahead of her to open the gate. She watched me calmly and strode through the gate as I opened it into the yard outside the calving barn. Shutting the gate, I gently prodded A48 into the barn and pulled the door shut behind her. After parking the four-wheeler next to the barn, I hustled inside, switched on the barn lights, and locked the heifer into a calving pen that had been freshly bedded earlier in the day. I decided to give her a few minutes to get back to giving birth while I checked with Marilyn and the rest of the family in our farmhouse.

Everybody was happy at home. They were waiting for me to join them for supper.

I sat down to hurriedly consume a hot roast beef sandwich with potatoes and gravy, cherry pie, and tea. It meant a lot to me that most of the ingredients on the supper table were produced on our farm—the roast beef, potatoes, cherries and homemade bread, although the flour and tea were not of our own producing.

"We might have to help A48," I suggested to Shelby and Jon. I wasn't sure which one of them might volunteer to help me. Shelby periodically assisted with birthing, which later in her life gave her a "leg up" when she was going through medical school training. After delivering her first human baby, her supervising obstetrician commented, "You handled that like a pro. How did you know what to do?"

"I helped my dad deliver calves," Shelby proclaimed triumphantly.

Tonight though, Shelby had a lot of homework to complete. She announced earlier in the week that she accepted an offer to attend the University of Colorado at Boulder, where I had completed my undergraduate training. She needed to finish applications for scholarships. Jon was occupied with practicing his drumming lessons and completing homework as well. So I slogged through the mud to the calving barn to check on A48 by myself. Flipping on the barn lights, I immediately concluded that A48 had made no progress giving birth. She was lying on the fresh straw, pushing hard, but a membrane stretched over the emerging calf's front feet and nose. "*She needs an episiotomy*," I surmised. Gladly, I thought, "*That's not an inheritable trait; she won't have to leave the herd.*"

I dashed to the house to collect a scalpel, antiseptic, and to rally someone to help me. Jon responded, saying he would be right out. Quickly, I filled a bucket half full of warm sudsy water into which I tossed the scalpel and a surgical needle. I stuffed a roll of surgical thread into my pocket. When I reached the barn again, A48 was standing with her back arched and pushing hard.

The heifer's troublesome membrane, her labia, stretched over the emerging calf's legs and nose. Shortly, Jon unlatched the barn door quietly so as to not disrupt A48's birthing process and joined me. "We'll have to tie her up," I proclaimed, reaching for the nearby lariat. I tossed the open loop over A48's head. She stood still, not offering any resistance, calmly watching Jon and me.

"If she's going to be cooperative, let's not restrain her," I submitted, while handing Jon the end of the rope in case he needed to cinch it around the tying post. I rinsed my hands in the still warm soapy water and carefully inspected A48's bottom.

"Hand me the scalpel." Jon thrust the knife into my hand.

Quickly, I made a ten-inch incision, starting where the labia attached to the heifer's vulva, then descending downward over the calf's nose and front feet. The labial membrane retracted. With both hands I grasped the calf's front legs and gave a light tug. The calf's head and chest popped out, up to its hip joints. A48 pushed as well. Another tug and a push and the calf flopped to the straw and gulped its first breath. Meanwhile the heifer stood stolidly as if anchored to the floor in the middle of the calving pen. Jon held the rope loosely a few feet from her head.

"Take the lariat off her head," I suggested to Jon. As soon as the rope was removed, A48 immediately swirled to inspect her calf. She vigorously sniffed her newborn prodigy on the straw. Instinctively, the young mother cow mooed softly and began licking a husky bull calf.

"Let's give them a few minutes to bond," I recommended. Jon and I sat down on bales of straw near the barn door.

"I've never seen a cow behave like that," Jon commented as we observed the young mother and her calf from our seated positions. "She acted like it didn't hurt when you cut her. It was like she knew we were trying to help."

"She's a first-calf heifer," I proclaimed. "I wouldn't expect an animal we have never worked with to be so calm either." Jon and I sat and gazed a few minutes, heavy in thought.

The calf raised and shook his head. Its mother lapped placenta and amniotic fluid from her calf's hair.

"I'll have to sew her up," I said. "Let's see if she will allow me to do that or if we will have to hitch her to the post."

If the episiotomy cut was small, the standard practice was to not repair the torn tissue, but this was a big incision. I strung three feet of surgical thread through the eye of the veterinary needle, grabbed a plastic jug of iodine solution with a squeeze trigger and approached A48's backside.

"Easy girl, I don't mean to hurt you." I sprayed iodine solution over her quivering rear anatomy and gently tucked the severed pieces into the cleft of her vulva, holding only the outside fragments of her labial membrane. Starting at the top I pushed the needle through both sides of the membrane, stitched them together and quickly made another dozen stitches, to hold the membrane together. I tied off the last knot, without the heifer shifting position except to move a foot or two to lick dry the various body parts of her baby.

"I've never seen anything like this before," I exulted while tossing the surgical instruments into the bucket.

"I thought she might have kicked us out of the barn by now," Jon uttered. "She certainly trusts us."

A few years later CNN reporter Judd Rose asked me how beef and dairy producers feel when they are forced to sell their cows. He was doing a story about how low cattle prices were causing some farmers and ranchers to sell their herds. Judd had already visited extensively with a farm couple who were forced to disperse their cattle and who had consulted me for the psychological depression they were dealing with. He interviewed me in the same barn where A48 had her first calf.

"Farm and ranch people feel vulnerable when the way they acquire meaning in their lives is threatened," I commented. "Livestock producers especially feel imperiled. The owners and the cows take care of each other, in different ways. It's a mutually beneficial relationship and one in which deep attachments develop."

"When farmers and ranchers are faced with the possibility of having to sell their cattle or land, they feel that not only are they letting themselves and their families down, they feel they are letting down the animals they've bonded with."

Judd Rose died of cancer a year later. I was told that this news piece was among those he requested for viewing at his memorial service.

Now, six years after I sold my last cow in 2004 because I have to be gone so much from home to lecture and teach around the US and other countries, a newspaper reporter asked me recently how I feel about that. "I still dream about my cows at least two or three times a week," I admitted. "I can still see all my cows. I know their names. I know their ear tags. They still mean a great deal to me."

Christmas Nugget

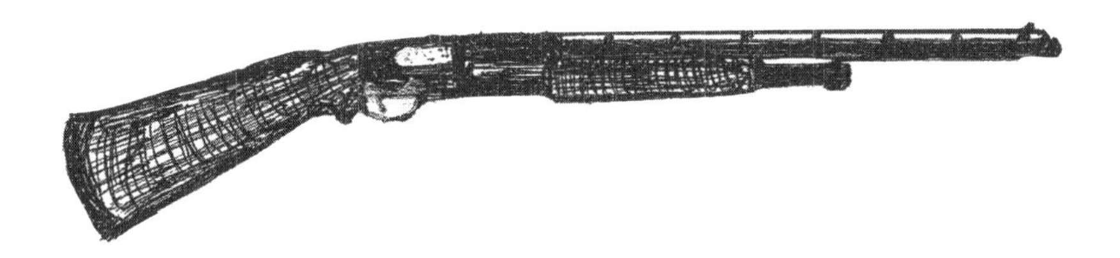

Most of December was cold, windy, and mean but okay for hunting pheasants, and the few ducks and geese that had not migrated to warmer climes.

Jon and I had planned to hunt and fish over the Christmas break. Would we be able to find any birds? We ruled out fishing because ice had started to form on the lakes and ponds but it was too thin to bear our weight. A skiff of snow dusted the surrounding farmland.

On Christmas Day, the eastern sky was lighting up without a cloud and there was little wind. By 6:45 AM Marilyn and I were already functioning, each in our own ways. Marilyn was making the spinach salad and a pecan pie that would accompany pheasant breasts in a Grand Marnier sauce that I prepared last evening for Marilyn to place in the oven later. I started the coffee. Even 26-year old Jon had arisen. His yellow Labrador retriever, Nugget, followed him from room to room, to make sure Jon didn't leave the house without her. We knew that she knew we would hunt today because she wagged her tail last night when the word "hunting" was mentioned while we cleaned our shotguns.

By sun-up the entire family had consumed a fit-for-a-king breakfast of Swiss toast, wonderful orange persimmons that made your mouth pucker, and Brazilian coffee. Jon, Nugget, and I headed to the fields while Marilyn pursued what, to her, were equally enjoyable pursuits of wrapping presents, calling friends on the phone, and playing Christmas music louder than I could stand. Jon and I left the house, knowing that most of the pheasants that had not yet been shot were in deep hiding.

Jon and I were as excited as seven-year-old Nugget as we tromped into the heavy prairie grass on my cousin's Conservation Reserve Program land. We had permission to hunt there and we hoped that we would have the entire 160 acres to ourselves while others who sometimes hunted there opened presents and stuffed themselves with Christmas dinner.

Two hours of brisk walking through the thick prairie grass yielded two hard-won pheasants. Nugget retrieved both birds promptly and proudly.

Close to noon, Jon and I reached a stretch of our farm that we hadn't hunted for several weeks. A creek ran through the middle of the dense filter strips on each side of the stream.

I like filter strips because the 120-foot wide stretches of prairie grasses along each side of the creek curtail runoff of fertilizer and manure into the waterway. The filter strips also furnish excellent cover for game birds, deer, and other wildlife.

Nearly noon, Jon and I hurried to traverse the quarter mile long stretches of filter strip before dinner time. Nugget did her best to push through thick prairie grass and reeds. Suddenly Nugget stopped, her tail wiggled side to side excitedly and her entire body shivered as she peered into a tangle of weeds and Indian grass.

"Nugget's onto something," Jon yelled. As Jon and I drew to within 30 feet of the site where Nugget was expectantly waiting, Jon commanded, "Get'em, Nugget." Eagerly the dog pounced into the tangle of weeds and grass. Two cackling cock pheasants winged upward, one heading north past Jon and the other heading west past me. Both our guns blasted simultaneously. Jon's pheasant tumbled first in a flurry of feathers and Nugget was on it in seconds. As she was retrieving Jon's rooster, my pheasant crumbled into a horseweed forest across the creek.

After Nugget deposited a brawny rooster next to Jon, she poised on her haunches at his left side. I wasn't sure if she realized another bird had fallen.

"Dead bird," Jon proclaimed as he pointed across the creek. Without hesitation, Nugget scurried down the 15 foot creek embankment, swam across the waterway and climbed the opposite bank. As Jon signaled Nugget to make a circle and commanded, "Find the bird, Nugget," the yellow Lab began circling the area to which Jon pointed. Within seconds she homed in on the downed pheasant. Seizing it, she swam back across the creek and presented the trophy to Jon. Just then, the carrilon pealed at the St. Boniface Church steeple two miles away in Westphalia. The air was quiet except for the church chimes sounding the notes of "*We Wish You a Merry Christmas.*" "It doesn't get any better than this," Jon commented as he stuffed the second pheasant in his vest. Nugget panted in agreement.

Two hours after a sumptuous Christmas dinner, Jon and I were back in the field. We tromped through mile-long stretches of brome and prairie grass filter strips on our neighbor Curt's farm. Jon and I had already shot five of our combined limit of six pheasants. I was tiring after a long day, so I stayed on the nearside of the creek while Jon and Nugget explored an island of weeds, woody shrubs, and grass a hundred yards away. While I was resting on my 12 gauge Winchester, a cock pheasant cackled and burst out of a patch of reed canary grass 70 yards ahead of me. It would be a long shot, I thought, and a bit tardy, as I swung my gun and fired twice. Several feathers drifted in the air. The wounded male pheasant winged another 300 yards along the grassed waterway ahead of me. When it landed in the heavy grass along the stream, a thought occurred to me, *"We'll never find that bird."*

Jon had turned to watch the spectacle of the distant pheasant flying onward after I shot it. He didn't believe me when I told him that I saw feathers fly. We stalked along the creek toward the site where the pheasant had alighted. Fortunately it was on the way to the road where our vehicle was parked, for we were all tired.

When we reached a point where I thought that the bird had descended to the ground, I asked Jon to send Nugget across the creek to my side. Nugget looked at her boss quizzically but she obeyed Jon's command. She scrambled across the creek and began surveying the prairie grass filter strip back and forth in search of a scent. Suddenly she began racing well ahead of us.

Surprised, Jon commented, "I think she's got a scent!" A quarter mile ahead, the grass through which Nugget was lurching abruptly stopped swaying. Jon cautiously spoke again, "I think she's found the bird." Jon and I waited hopefully.

Ten minutes later, Nugget returned with a large male pheasant in her mouth. This time she deposited the bird at my feet.

I patted her on the head. "This is a Christmas nugget I'll always remember."

"No better Christmas than this," Jon volunteered, and I nodded. Nugget smiled too as she panted.

As my brother, Larry, used to declare, "Excellent joy for all!"

In the Blink of an Eye

"There's got to be something you can do," the balding 50ish man implored me. "I know there's more going on in his head than he can get out. He has a lot of stuff going on inside his brain but he can't tell us what he's thinking."

The portly gentleman's big belly quivered under his tightly stretched white t-shirt. He plucked a tear away from his right eye with a thick forefinger. I turned my head aside briefly to give him a chance to regain his composure.

"Don't you have some kind of psychological test that you can give him, Doc? He just lies in his bed all day long or the nurses tie him in his wheelchair and park it in front of the TV so he can watch it in the recreation room."

Joe was telling me about his 29-year-old quadriplegic son, Steve, who was involved in a tragic auto crash while he was driving home from work several years earlier. A teenage driver, impaired by alcohol, steered her automobile into the path of Steve's oncoming car on a two-lane highway, eight miles south of town. The accident happened at 10:30 at night. No one heard the blare of Steve's car horn and the horrible clash of the two vehicles on the rural highway. Several minutes lapsed until a passerby came onto the scene and dialed 9-1-1 to summon help.

Now, five years later and after a lawsuit, Joe still had no good answers. Steve had tried to steer his vehicle onto the highway's right shoulder but the drunk teen's car turned too quickly into his automobile. Her car smashed into the driver's side of Steve's vehicle. The terrible impact whipped Steve's head back and forth, severing his spinal cord just below the base of his skull. The left side of his head was smashed in and much of the left side of his brain was severely bruised beyond the point of further use. Steve couldn't talk. He regained partial use of his throat so that he could swallow but the rest of his body was limp and out of his control. After many months of physical therapy he was deemed to have achieved maximum treatment benefits. Steve was now in his third year of residence at St. Mary's Nursing Home.

Passionately, Joe went on, "There's got to be a way you can show Vocational Rehab. that Steve understands and wants to communicate. We'll never get a settlement out of the Court, so we need Voc. Rehab. to step in. I know he can hear me because his eyes light up when I say something he likes. We ought to be able to find the equipment he needs. See what you can do, Doc."

Five days later I entered the lobby of St. Mary's Nursing Home, carrying the Peabody Picture Vocabulary Test with me. I hoped I could get Steve to show me which of four pictures on a page of the exam book represented a word that I announced.

I had not met Steve before. I didn't know if my plan would work. After greeting the supervisor at the nurse's station, she showed me to Steve's room. I smiled and said hello to Steve as I grasped his limp right hand. Steve had no reaction that I could notice. The nurse raised the bed so Steve's head and upper body were partially elevated.

Steve looked at me quizzically when I said, "I'm Dr. Rosmann. I'm a psychologist. Your dad asked me to visit you."

Steve's eyes seemed to darken. Had I offended him? Then I realized that perhaps I'd inferred his father was concerned that Steve had a mental health problem. I added, "Your father thinks that if we can figure out a way for you to demonstrate how capable you are, perhaps we can explore some assistance from Vocational Rehabilitation Services. We have to figure out how to show them what you understand and want to communicate."

Steve's eyes brightened, just like his dad said they would.

"Do you understand what I'm saying," I questioned. Steve's gaze remained fixed on me, unchanged but bright.

For a few moments I pondered what to do. *"Please give me a hand here,"* I prayed silently.

Then a thought came to me, "If you hear me, blink your eyes, Steve."

Steve blinked.

"Great," I announced happily. "Did you understand what I said earlier about trying to show how much you understand and want to communicate?"

Again, Steve blinked.

"This is so good," I blurted joyfully. "Let's show them!" I briefly explained the directions to Steve about how I would present him four pictures on a page while pronouncing a single word. Then I would point to each picture. I asked Steve to blink when I pointed to the correct picture that illustrated the word I announced.

We started with simple words, at the third grade level. It was immediately apparent from Steve's eager blinks that these words and concepts were too simple. Quickly, we advanced into the high school range and above. There were some concepts that I expected Steve to know that puzzled him, and left me wondering why he didn't recognize them. Perhaps the brain tissue that had earlier incorporated the meaning of these particular concepts had been destroyed by the trauma to his brain.

I wanted to see if Steve could relearn complicated vocabulary words. I repeated some of the items that he had missed earlier. He answered nearly all of them correctly.

Eventually we reached a level of difficulty where Steve regularly missed all the test items. It was no use proceeding any higher. When I scored the results, Steve's measured intelligence quotient was 86, in the low average range. I explained that a score of 86 was encouraging, given his brain trauma. It was also hopeful that Steve demonstrated the capacity to retain new information. Both were signs that might convince Vocational Rehabilitation officials to work with Steve. "I don't want to get your hopes up that this will necessarily take place," I admitted, "but I hope that the people who make these kinds of decisions can equip you with technology that you can control with your eye movements." I had seen how proficient the physicist, Dr. Stephen Hawking, was using similar technology. Impaired by amyotrophic lateral sclerosis to the point that he could not speak or type, Dr. Hawking continued to write books that advanced the knowledge of the formation of our universe. It was finally Steve's turn to have something good happen to him.

A year or so later, now January 1994, I stood next to my grandmother's bed at Elm Crest Nursing Home, in Harlan, Iowa. Grandma was dying. She had been hospitalized a couple weeks earlier with a severe bladder infection. The doctors and the hospital staff did everything they could to help Katie overcome her infection, but Grandma was getting weaker still.

My mother, who was Grandma's oldest daughter, and I visited her daily at the hospital and conferred with her caretakers. Grandma told us that she didn't want any special life-saving means taken to keep her alive. She had announced for many years that, "I don't know why the good Lord keeps me alive. I'm ready to go any time He wants me. I look forward to seeing Henry (her deceased husband) again." Grandma was 98 years old. Her living will indicated her wishes.

When Grandma's bacteria count in her urine had dropped to an acceptable level, the hospital discharged her back to Elm Crest. She continued to take antibiotics orally but her body was failing. Grandma was so weak she could not sit up. She sipped water and a little juice but she had trouble swallowing solids, so these were not pushed by the nursing home staff or by Mom and me when we visited her. Her other children who lived within driving distance came by and several other grandchildren visited her frequently. Her Catholic pastor was summoned to administer the last rites.

Over the following few days Grandma's breathing became increasingly shallow and at times irregular. I stayed sometimes for an hour with Grandma alone or for shorter durations when others were also visiting.

We held her hand, prayed the rosary, and talked quietly and reverently about accomplishments in Grandma's life that we savored.

Grandma was the unofficial poet laureate of Shelby County. In prior years she would boldly hold her right hand in the air, forefinger pointed skyward, and declare, "When I was a girl," and go on to relate an interesting experience or observation that had occurred to her. Each story related a simple—or sometimes profound—nuance of life that inspired listeners. She rode in the dignitaries' convertible car when her town, Defiance, Iowa, held its centennial parade a few years earlier.

The last time I saw Grandma alive, she was different. When I entered her Elm Crest room, empty of visitors other than me, she lay huddled under mounds of blankets, barely breathing. "Hi Grandma," I proclaimed. She gave no response. "It's kind of cold out," I pronounced in a way that I hoped would stir a reaction, but she registered no sign of understanding me. Grandma was a farm woman all her life and usually paid much attention to the weather and crop-growing conditions. I reached under the covers to hold Grandma's hand. It was cool.

"You might not have much longer to live," I commented. I had been with several people, persons I had worked with as a psychologist, to help them prepare for death. I was fortunate enough to be around when they expired. I knew what was taking place with Grandma.

"Are you ready to leave this life?" Grandma feebly winced and seemed to try to move her mouth as if wanting to speak. No sound came forth.

"If you can hear me, Grandma, blink your eyes." Her eyes had been shut the whole time, but now they cracked open and Grandma discernibly blinked them shut.

"Are you scared?"

Grandma blinked again. The muscles on her face quivered slightly. She seemed to try to work her jaw. "I didn't get that right, did I Grandma? You're not scared, but something else is bothering you, right!"

Grandma blinked affirmatively.

"What is it, Grandma? I know you are ready to go." I paused, and then uttered a thought that came to me, "Are you worried about the rest of us, like my Mom and your family?" Grandma blinked assertively.

"You can help us more in our lives here when you get to your next life," I suggested, as I squeezed her hand.

Grandma's face relaxed. "Would you like to pray Grandma?" Grandma's eyelids parted and shut again.

"Our Father, who art in heaven, hallowed be Thy name… ," I intoned. "Dear Father, please give Grandma safe passing."

Grandma's breathing was barely discernible.

A few minutes passed, "Are you ready to go now?"

Grandma's eyes barely cracked and calmly creased shut. "Goodbye Grandma. I'm going to get Mom."

Despite tears clouding my eyes, excellent joy surged through my entire being as I hastened to my mother's apartment, attached to the Elm Crest Nursing Home. My mother was getting fairly old herself and had taken up residence in the Elm Crest community several years ago.

The thought crossed my mind, "*Funny how life comes and goes in the blink of an eye.*"

Nothing Wrong With Being Second

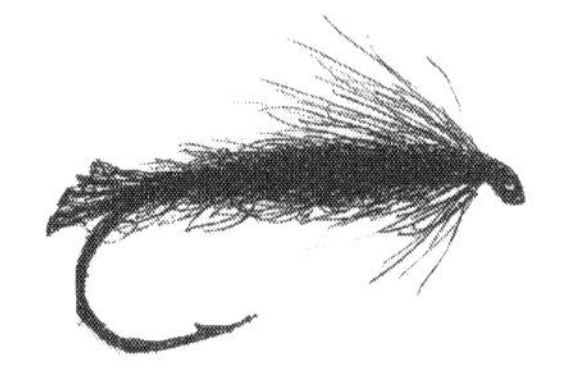

Jon and I scrambled to stuff fishing pontoons, spare float tubes, a large tent, folding chairs, cots to sleep on, coolers, and fishing gear into the ten-foot trailer. Our jeep was already packed full of items we would need to fish the Norfolk and White rivers in northern Arkansas. We timed our venture with the start of major league baseball season.

Timing was crucial. Fishing the Norfolk and White rivers was best before the trees fully leafed out and before hoards of fishers descended onto these well known waters. If one or more electrical generators were running on the Norfolk Dam, the water would flow briskly and fishing for brown and rainbow trout and occasional smallmouth bass could be challenging but if the water flow at the dam was turned off, fishing the Norfolk River could be really hot. We could begin floating the Norfolk River at the campground just below the Norfolk Dam and we could exit the waters at the parking lot next to the *"Handicap Accessible Deck"* overlooking the Norfolk River or farther downstream at the confluence of the Norfolk and White rivers.

Jon had figured out the solution to successful fishing during the two previous seasons and landed one nice trout after another, mostly browns, on orange sand worms, micro jigs, and tiny zebra midges that he purchased from any of several fly shops. Most of the bait shops offered transportation services so one could park the trailer and jeep at the take-out point and the transporter would ferry the eager fishermen back to the campgrounds where they would put their pontoons into the river.

We became friends with Jeannie over the years. She cried when she announced to us that she was closing her business to look for new opportunities. She had shared most of the gritty details of her divorce, drinking escapades, and the death of her brother who had started the bait shop and transportation service several years ago. She seemed to have come to terms with her life. We liked her pleasant manner. Unlike some proprietors, Jeannie never overcharged us.

Jim, a fairly famous fly tier and guide who Jon engaged, was much more taciturn. Featured in numerous articles that appeared in fly fishing magazines, he was a master fisherman, probably of the same rank as Walt, but much more critical. I figured out that his patience would run out if the student didn't master his techniques within two tries. Disgustedly, he would take the rod from his student, scold him to pay more attention and repeat his instructions only one more time. Thereafter, you were on your own!

I supposed Jim was fed up with nincompoops who showed up at his fly shop with fishing equipment worth many thousands of dollars, proclaiming they were experts, only to find out they couldn't cast a fly twenty feet.

Last year Jim took Jon and me to the White River just below Bull Shoals Dam to fish a stretch limited to catch and release fly fishing. The water was crystal clear and spread out nearly evenly across expanses 100 feet wide and two feet or less deep. I couldn't see any fish in the water but Jim could. I had never fished this kind of water before. Jim gave us some helpful lessons on fly line mending and minimizing backcasts. He also supplied plenty of his own expensive micro jigs and zebra midge flies, but no lunch, only water, and he expected us to appreciate his style of fishing as much as he did. Catching and releasing all trout seemed okay to Jim. He said he didn't like to eat trout. Jon and I relished charcoal roasted trout stuffed with peppers, onions, tomatoes, and fresh herbs. However, I learned a few things, probably a whole lot more than Jim suspected.

Jim tried to get me to change my method of retrieving fly line. I had learned a method of using only my left hand to retract fly line and store it between my left thumb and forefinger from an article in a fly fishing magazine. I started to explain that my father-in-law thought this was a more effective way of retrieving line than the standard method of pulling fly line with one's left hand through the thumb and forefinger of the right hand, while also holding the rod in one's right hand, but Jim cut off my explanation. Jim also chastised me for cranking my reel with my right hand and insisted that I use one of his rods, with the crank on the left side. I started to explain that I owned a couple fly reels that cranked on the left side but when I saw that Jim wasn't interested, I didn't finish my comments. Using a right-handed reel felt more comfortable to me so I returned Jim's rod to him.

Jim attached sticky, yellow, round strike indicators to the leader about eight feet above the fly, something Jon and I had tried a couple times earlier and dismissed. Jim fished nearly entirely by using visual cues rather than tactual cues. One had to watch when the indicator disappeared below the surface of the water, for it was then time to set the hook, rather than to wait until you could feel the tip of the rod sink. Jim made sure I figured out that I was missing quite a few fish that had briefly mouthed the fly and discarded it before I could set the hook because I was accustomed to tactual cues. Jon, with a younger brain and more malleable behavior, caught on faster than me. Jon captured several eighteen-inch rainbows and reaped abundant praise from Jim while I earned only an occasional hook-up and mostly disgusted glances and chastising comments from Jim. I became a bit demoralized.

Gradually stepping ever farther downstream from Jim and Jon, I cast into the swiftly flowing, mostly smooth water until I was well beyond voice range of Jim. Then I stepped even more hastily downstream until I reached a secluded spot below a small island where ripples on either side of the island reached a confluence that had scooped out a five-foot deep depression.

I glanced upstream to see if Jim was watching me. Jim seemed completely absorbed in helping Jon release another fine rainbow. I remembered Jim's last comment, "Never use your left hand to hold the fly line. You can't set the hook quickly and you can't bring in your line fast enough when you hook a fish."

Turning my back toward Jim and Jon, I reached into my fishing vest to bring out a round, clear plastic, electrical tape container that held flies I had tied myself. Searching through the flies I located a size 14 nymph with an olive green mohair body, a red wool tail, and brown hackle behind the head. I snipped off the foam yellow strike indicator. With a single backcast and forward thrust, learned from Jim earlier, I cast the previously untried wet fly into the rivulet nearest me as it coursed around the small stony island just upstream. I let the fly sink to the cobblestone bottom as it rushed toward the hole below the island. Using my left hand, I gradually retracted fly line and stored coiled line between my left thumb and forefinger in the fashion most comfortable to me. I noticed the leader and tip of the fly line start to straighten out, even though I could feel no additional tension on the tip of the rod. I thrust the fly rod upward and held on to the fly line in my left hand. A huge rainbow trout, at least two feet long, burst out of the bubbles in the pool. It splashed hard, zigzagged across the pool and jerked coils of fly line from between my left forefinger and thumb. The battle seesawed back and forth for a couple minutes while I nervously glanced upstream to see if Jon or Jim were watching and following the movements of the grand trout surging ahead of me.

As minutes passed I maneuvered the big rainbow toward my outstretched net. Raising the fishnet I enveloped the huge fish inside. Turning my back toward my companions upstream, I leaned over, as if checking my fly. With the fly line and tippet now relaxed, the red-butted green nymph fly fell out of the trout's mouth without any assistance. I dipped the net into the water and the huge fish swam away. Quickly I turned and cast again into the same pool.

A half hour later, Jon, Jim, and I joined together to call it a day. Jon and I were hungry, for it was 4 PM and we had not consumed any food since breakfast. Jim offered each of us a bottle of water and I accepted, adding that we would make supper back at campsite later. Jim asked, "Did you have fun today?"

"It was great," Jon responded.

"Yeah I learned a lot," I said. While looking somewhat askance at Jim, I continued, "but, I'm sure I only got second place today."

As Jon and I were driving back to our campsite, we were both pleased as we talked over what we had learned today. I told Jon about the big fish I had caught using a combination of new knowledge from Jim and my own tried and tested methods of fly fishing. We were both looking forward to substantial suppers of trout we had caught yesterday, cooked over a charcoal fire, preceded by renewing our acquaintance with Jack Daniels.

On the way back to our campsite, Jon drove the Jeep through Flippin, Arkansas. On the outskirts of town, we spied a sign at the entrance of a church that said, "Welcome to the Flippin First Church of the Gospel Truth."

That got us revved up. We had noticed that, as we were driving along, nearly always there was a liquor store within 300 yards of a church. Jon, who had finished a business major at the University of Iowa, suggested that there might be an entrepreneurial opportunity here. "We'll create the Second Church of the Gospel Truth and Liquor Emporium."

"Yeah, we don't want to be first!" I coaxed. "What's wrong with being second? We can't all be first. The Second Church and Liquor Emporium could be a real winner."

"One stop satisfies whatever ails you," Jon laughed.

"Tank 'em on Saturday night and spank 'em on Sunday," I rejoined. We were both mightily satisfied that with all the seconds, thirds, and "later-ons" out there, the Second Church of the Gospel Truth and Liquor Emporium could be a sure thing.

After a sumptuous supper, conversations with Jack Daniels, and solid sleep, Jon and I were ready for the next day. We arose in the morning to find that all the generators had been turned off. It was a no-brainer to float the Norfolk River. Hastily, I called Jeannie to take our Jeep and trailer down to the take-out point. We were on the water by 8 AM

Hardly had we departed from the Norfolk River shore next to the campground when Jon almost immediately hooked up with a 17-inch rainbow. He fought the fish gallantly and landed it before I floated downstream and out of sight. I watched my son observantly and was glad that Jon was successful and having fun.

After an hour and a half of slow fishing with few bites, I began to wonder if I should change flies. I had started fishing with zebra midges, then micro jigs and finally tried red-orange sand worms that I had purchased from Jim the day before. I had fair success using these flies two days earlier, but not nearly the good fortune that Jon experienced. Jon caught 30 fish and released all but his limit of five while I caught a dozen and released all but two fish.

I wondered what flies to use. After lighting up a cigar, I tied on the same red butted green nymph that had worked yesterday when I escaped from Jim's tutelage. After a half hour, I had three hits and one fish but not the kind of action I was accustomed to. It was time to try something else.

Over the next couple hours, I tried the Big Shrimp Special, numerous nymph patterns, and some streamers I had tied. Here and there, I had a hit or two but it was slow going. Then I remembered a fly I hadn't tried for a long time.

A few years earlier, Marilyn and I celebrated our 25th wedding anniversary in the British Isles. A wonderful trip, we spent time in Ireland, England, and Scotland. We sought out the Lake Country in northwest England where Beatrix Potter had lived and authored her many children's books. The guest house where we stayed had been her home. Hoping that there might be an opportunity to fish in the British Isles, I had packed my fishing vest, a couple reels, and a few boxes of my most treasured flies.

The morning after we arrived at Lake Windermere, we rented a boat and a fishing rod. I tried several fly patterns that I had brought with me.

No hits on the first flies that I tried. Then I tied on a green wooly worm with a brass bead head and two split tails made from goose quill. Almost immediately I hooked a 15-inch brown trout. Reeling the brown trout to the boat, I was especially pleased that I had now caught trout on two continents. I started to deposit the still-wiggling fish onto the chain stringer that I brought to England when I remembered that the merchant who had rented us the boat had given me a priest. Curious, I had asked what the priest was for. It is a foot-long handle sawed from an oar. The proprietor explained how to use the priest: "When you catch a fish, you administer the last rites by hitting it over the head." I did just that.

Marilyn and I motored along the shoreline of Lake Windermere for a couple more hours. Marilyn steered the motor on the boat while I cast into likely spots. I caught two more similar sized trout on the same fly.

That evening we presented the trout to the chef of the Lake Windermere Lodge. The trout the chef prepared was one of the best culinary delights that Marilyn and I ever experienced. The chef came to our table to thank us for presenting the trout to him and commented, "These are the second best trout that I have ever had the honor of preparing. The best trout I ever prepared were for the Queen of England."

As I now regaled in the pleasant thoughts of the wedding anniversary trip I tied on a green wooly worm with the brass bead head and two goose quill tails. I cast into the deep recesses of slow moving water along the shore 40 feet to my right. As my pontoon slowly drifted downstream, the fly line followed. I held loose fly line in my left hand, between my thumb and forefinger. I felt a trifle guilty as I remembered that Jim told me to hold the rod in the palm of my right hand and to bring the fly line through my right thumb and forefinger next to the reel. It still felt uncomfortable following the instructions that I had been given the day before so instead I adhered to the old method that I had learned 37 years earlier on my own and with Walt's blessing.

Shortly, there was a solid tug on the line stretched between my left thumb and forefinger. Automatically I raised the rod tip. A husky speckled trout exposed its back at the surface of the Norfolk River 45 feet in front of me. I tried to steer the sturdily swimming fish toward me but the trout stubbornly pushed downstream. Gradually, I turned the trout so it could no longer use the flowing water to assist its intended course. I maneuvered the fish toward the evermore shallow shoreline where I stood up in the pontoon and finally I scooped an 18-inch brown trout into my net.

I spent another hour in the same spot and landed three more equally nice fish while releasing several smaller rainbow and brown trout. Then I pushed my pontoon into the free-flowing stretch and meandered down the Norfolk River.

Three hours later, I arrived at the *"Handicap Accessible"* parking lot. I pulled onto the shore, just as Jon came around a riverbend 50 yards upstream. Jon shouted, "Dad, how many did you catch?"

"I kept four and let quite a few go," I responded. "How did you do?"

"I think I caught at least a hundred," Jon responded. "But I only kept five, so we're okay."

As Jon paddled to the shore next to me, he volunteered, "I caught them all on the green micro jig."

Not to be outdone, I quietly uttered, "I caught most of mine on flies that I tied. They worked better than the flies we bought from Jim." To myself I thought, *"Nothing wrong with being second!"*

It's What Walt Would Have Wanted

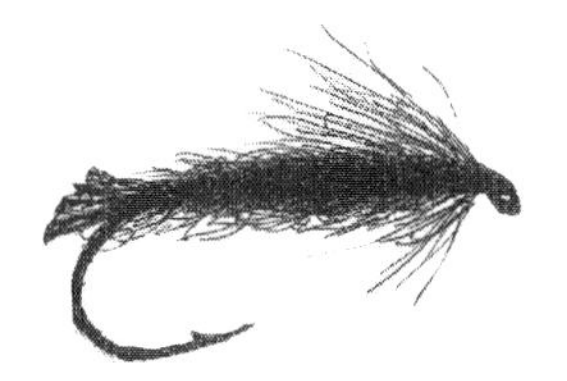

Marilyn called on Friday night to tell me that her mother, Michi, had died that afternoon at the Burley, Idaho, hospital. Marilyn was at her bedside when Michi passed away, for Marilyn had hurriedly jetted to Idaho from the Midwest six days earlier when she learned that her mother had suffered a massive brain hemorrhage. Walt, her father, had died only nine months earlier.

Walt was an inspiration for many fly fishermen and fly-tiers in the Burley area, many of whom attended his funeral. Walt fished at least twice weekly for 40 years and more often when it was opportune for him to avoid Michi's "honey-do" lists. Walt and I formed a close bond when fishing together and Jon, my son, made it a happy threesome when he was old enough to join us. I telephoned Jon, now 26, and we made plans to attend Grandma's funeral.

Marilyn gave me orders to, "bring the truck and the trailer out so we can take some things back home when we start cleaning out the house." Just what I wanted—more things in the basement and the garage to take up space that fishing equipment could occupy. Jon and I knew better though than to disobey the order.

We liked Grandma. She was pretty cool about fishing and she had never made a fuss when we cleaned fish in the kitchen sink. Mothers-in-laws don't get much better than that. Besides that, she did a pretty good job raising Marilyn. Marilyn came into our marriage with marvelous tolerance for hunting, fishing, and eating wild game, so a few minor imperfections, such as stocking the basement and the garage with things you don't need, could be overlooked.

Marilyn also advised me to, "bring only the clothes and a few things you need because all the space in the truck and trailer will be taken up with things to bring home." What a wonderful phrase: "Bring only the clothes and a few things you need."

Jon and I packed our clothes and personal kits sparsely. Then we packed what we needed: two fishing rods apiece, waders, float tubes, fish bags, fins, nets, a large cooler, cigars, warm underwear because it was October and getting cold in Idaho, and lots of Walt's hand-tied flies. We set out from western Iowa at 5 PM on Sunday evening to drive the 1,200 miles to Burley, Idaho.

Taking turns at the wheel, by 8 AM the next morning Jon and I rolled up to a favorite reservoir in southern Idaho where we often fished with Walt. It took only 16 hours (you gain an hour crossing into Mountain Time) to cover 1,200 miles. The only problem was that we didn't have Idaho fishing licenses and no fishing business establishments were yet open in the towns closest to the reservoir. To make matters worse, it was the first day of deer hunting season and the game wardens were out in force. "Hopefully," we rationalized, "the game wardens will be checking hunters instead of fishermen."

Hurriedly, Jon and I pulled on warm clothing to wear under our waders. The cold we were concerned about wasn't the air temperature—but the reservoir, which was 5,400 feet above sea level. We sorted through Walt's flies and tied on likely suspects to entice trout. Hardly had I paddled 100 feet from the reservoir shore when I had my first hit. What a beauty! Five minutes later I landed a 21-inch brown trout. Within minutes Jon had his first catch—an 18-inch rainbow trout. The sun shone gloriously on us. The cool morning air heated to 55° F within two hours. By then, and two cigars later, Jon and I had our limits and the smallest keeper was 17 inches.

One might wonder why worry about a limit when you don't have a fishing license. Jon and I wondered about the same thing. Each time a vehicle wound its way up the gravel road past the reservoir we grew a little more worried. Practically every brown pickup truck had an Idaho Fish and Game insignia on the door panels. Ten fish at $100 apiece could add up quickly into a considerable fine. "We better quit," I mentioned. "We haven't checked in with Mom by cell phone for awhile and she's probably wondering where we are."

As we paddled to the shoreline, Jon ventured that probably Mom thought we were driving in the mountains and telephone reception would be poor so she wouldn't worry much. Besides, she probably wouldn't expect us until at least noon. We wouldn't tell her we had gone fishing unless it was absolutely necessary.

Quickly we tore off our fishing gear, threw it in the truck and skedaddled. We arrived in Burley at noon. As we pulled up to park curbside in front of the house, Marilyn came outside to welcome us. "Wow," she said, "you guys made really good time."

"Yeah, it was great, no problems or anything," Jon ventured as he and his mom hugged.

"How come you didn't call me this morning?" Marilyn asked.

"We were in the mountains and reception wasn't very good," I responded.

Marilyn paused and then demurely probed, "You guys went fishing, didn't you?"

Jon and I boldly gave the answer we had practiced many times over the past 18 hours, "It's what Walt would have wanted."

Walt must have been looking down on us today and maybe Marilyn's mother too. After Marilyn saw the fish and heard our excuses for not being able to purchase fishing licenses, she offered, "You're probably right."

I volunteered, "We only brought what you told us—just a few clothes and the things we really needed." Marilyn's brother, Don, who had come outside to greet us too, smiled a bit jealously I think. Don was just getting into fishing and hunting over the past few years and had missed out on the wonderful experiences we enjoyed with Walt. But ... it's never too late to get into these pursuits.

Walt and Michi must have been looking down on Jon and me the next day too. Marilyn told us that we didn't need to clean out the basement that morning because she and her brothers were going to meet with the attorney to go over the last will and testament. Marilyn didn't want Jon and me loading the truck and trailer without her supervision.

Jon and I needed no more urging. Three hours later, we returned with five trout apiece, none less than 18 inches and this time all legally caught on two-day fishing permits that listed the first day as yesterday. It had taken considerable explaining to convince the clerk who sold us the licenses that yesterday was already past but, "It's what Walt would have wanted."

Slice of American Pie

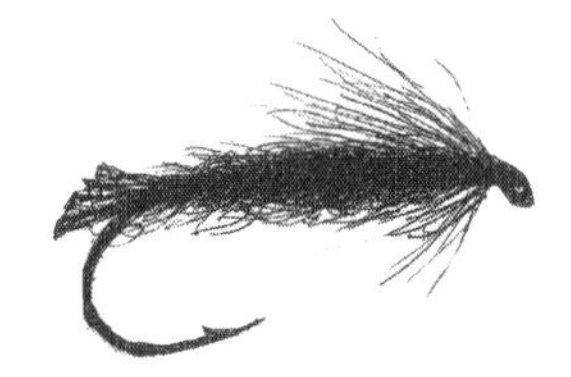

Eager to dangle our newly purchased light sensitive Swedish pimples in the icy waters of Mille Lacs, Jon and I motored up and down the snow-packed streets of Sunset Bay, Minnesota, to locate the proprietor who managed the string of cabins that included the one Jon had rented for three days of fishing bliss. It was late Friday afternoon in mid-January. The outside air temperature was -11 F. The ice was 13 inches thick, enough to support one-ton trucks that were heavier than our Jeep even though it was crammed full of camping gear, fish finders, insulated coveralls, and plenty of stuff we probably wouldn't need. Jon had thoughtfully advised Mom before we left home that she didn't have to worry about the ice being thick enough to drive on, and I must admit I was a tad bit relieved too. Plenty of others had concluded the ice was sufficient to bear them for there were many villages of shacks, trailers, and vehicles of all sorts scattered over the lake, several thousand in all.

Of course we had made the perfunctory stop at Cabela's to purchase our licenses and the newest gizmos for locating and attracting fish. Jon had linked us up with a highly recommended guide, Corey, for the next day. Quickly, we located the proprietor and got our cabin key. We drove several miles out onto the ice and soon found our cabin. We were ready for action!

That first night in the 8 x 10 ft. fishing cabin was anything but blissful though. The shack had five already drilled holes located inside the cabin around a wobbly card table and a couple ancient wooden chairs. Two, 2 x 6 ft. bunk beds—stacked one above the other, and a propane-fed stove were the only other accoutrements. The ice in the holes was 8 inches thick so Jon cranked up the motorized ice auger to clean out the holes. Eye-watering exhaust promptly inundated the cabin so we opened the door to allow fresh air to infiltrate. The pilot light on the stove blew out and we had to restart the stove a couple times. Only a great supper of Iowa beef steaks, thoughts of walleyes for breakfast, and plenty of advice from Jack Daniels restored our perspective.

While Jack hastened the first couple hours of sleep, worries about the propane causing carbon monoxide poisoning and arms flopping uncomfortably over the bedsides because their narrow construction wasn't designed for wide farmers' shoulders awakened me by 2:00 AM. Jon awoke cussing a short while later. Stratification of the cabin air made his top bunk too damned hot while a cold breeze circulated around me on the bottom.

We checked and rechecked the Swedish pimples, minnows and wax worms on the rods and bounced the baits off the lake-bed to no avail. We talked and compared life stories. Periodically other sleepless fishers rambled by in their pick-up trucks, causing the ice to crack loudly and menacingly. Whenever we went outside to pee on the snow that was piled alongside the cabin sills we uncovered mounds of excrement from previous occupants. Well before breakfast, at daybreak, we concluded this whole arrangement was a piece of shit. We weren't going to stay another night in this crappy place, even though Jon had pre-purchased the package as my Christmas gift.

After throwing our belongings into the Jeep Jon and I drove back to Sunset Bay to check out alternative accommodations. We located Pete's Bar and Grill, a white clapboard restaurant with a conglomeration of trucks, SUVs, and snowmobiles parked outside, where we were to meet Corey. The smoky building meant for about 60 people was already crowded with at least 100 patrons gobbling hefty plates of breakfast. It took only a few seconds to spot Pete, the head of this operation, for he reigned supreme at the open end of a horseshoe shaped bar already congested with coffee cups, half filled ashtrays, and whisky smelling glasses. The place was crowded with big, well-insulated guys, and a couple equally hefty women all chowing down toast, eggs, and sausage. Sidling up to the only open barstool and plopping down, I asked Pete if he knew of anyone who might have an on-shore heated cabin with a bathroom and shower that we could rent for a couple nights. Pete gestured for his wife to take over the management of the establishment and pulled Jon and me into a quiet space in the kitchen behind the bar. "Yah, I think I got a room because a guy didn't show up last night—at least his key is still here. We can check it out."

Exiting the bar and grill we followed Pete and his faithful Springer Spangiel, Matty, to the vestibule of a nearby church a block away. "You can stay here if the guy didn't use the room," Pete announced as he thrust a key into the door of the inner sanctuary. The converted sanctuary was divided into two large, warm, high-ceiling rooms. The room Pete showed us obviously had not been used so Pete shoved the key into my hand and said, "It's yours."

Although lacking any ornaments, the room was as functional as Pete. No drapes—they weren't needed because the church windows were frosted. There were two queen-sized beds, cable TV, and a big bathroom with a shower and two king-sized towels. "We'll take it," I pronounced. "Do you know Corey, because we're supposed to meet him at your bar and grill?"

"Of course," Pete responded. "He's a really nice guy and a very good guide. You'll get your money's worth."

As the three of us guys and Matty, all pleased with the way things were going, sauntered back to Pete's Bar and Grill, I asked Pete how he acquired the church.

"Apparently I bought it last year when I was too drunk to know better," Pete replied. "The next day the fellow I bought it from told me he hoped I would make good on the deal. He looked kinda sad when he said he was the church Elder, but he said the congregation had been trying to sell it for quite some time. Anyhow, it's been a good deal 'cause I got it real cheap and it didn't take a whole lot to turn it into two rental units."

I thought about suggesting that perhaps, *"more goes on now in the two motel rooms than went on when it was a church,"* but then I thought the better of it and decided to keep my cynical religious deliberations to myself. Jon changed the subject as he asked Pete what kind of fishing rods people were using here. Pete said he would let Jon use his new rod for a couple days. Entering the bar and grill, Pete pointed out Corey to Jon and me.

Corey was a clean-cut guy in his mid-30s with insulated coveralls. He sat next to his similar appearing accomplice, Dean. Corey asked if we had a GPS, which Jon had, of course. He told us to follow in our Jeep a couple hundred feed behind their Ford F150 and to record the GPS coordinates. We made our way many miles and at least forty minutes before Corey finally parked in a desolate spot with no other people in sight. Jon and I were getting pumped up.

Quickly, Corey and Dean drilled holes in the ice while Jon and I rigged up our rods. Corey explained that there was a small but discernable ridge somewhere under us. He pointed out the remnants of drill holes he and Dean had poked in the ice a few days earlier. Walleyes and possibly perch hung around the structure in the lake bed. He said we had come a long ways onto Mille Lacs because this part of the lake had not been fished hard and there was a good chance of catching nice fish. The song, *"Anticipation,"* started playing in my head.

Several hours later as dusk fell, the four of us had filled a five-gallon bucket with walleyes and a few perch. Jon landed a big-bellied walleye just shy of the 28-inch slot limit that Corey proclaimed would earn Jon a photograph in the *Sunset Bay Newsletter*. We were pleased and hungry for supper.

The night was pitch black except for a sky full of stars and the town lights when we arrived back at Pete's. We decided to take up the offer in Pete's brochure—*"bring in the fish and we'll clean 'em and cook 'em."* When I strode into Pete's Bar and Grill with the bucket full of fish a chorus of *oohs* and *ahs* arose. I haven't experienced so many backslaps and congratulations for a long time. Everybody wanted to see the fish and know what we caught them on. Corey and Dean puffed out their chests and displayed the photograph of the near-28-inch walleye on Corey's electronic camera. No one else in the establishment had caught even half as many fish. Pete rang his cowbell to offer a round of free drinks for all. He also grabbed the bucket and found two fellows who agreed to butcher the fish. Many patrons at Pete's ate well that night!

The next day was Sunday. It was doubtful that many local people went to church at Sunset Bay anymore, or maybe they didn't attend church previously and that is why the Elder sold the building to Pete so cheaply. Then again, maybe folks just got more out of fishing. Like my father-in-law and I, the hardy Minnesotans might have experienced reverence for God through spending time outdoors. Walt often told me as we drove to and from fishing haunts that he felt closer to God when he was in his float tube, or while wading his favorite central Idaho stream, than he felt sitting in church. I could relate to that, for I have often felt more intense closeness to God while meditating in a hunting blind, floating a river, or working in my farm fields than in a church pew.

Jon and I heated up leftover walleyes from last night's supper on our camp stove for breakfast and soon headed out to the same spot on Mille Lacs where we fished yesterday. We wanted to get there in time for the mid-morning surge of bites. The day was cold and windy. It felt good to have an ice-fishing hut with a portable heater to offset the -13 F degree outside temperature. Hard telling how cold the wind chill was.

Seven hours and several cigars later we called it quits. Even though we didn't catch as many fish as yesterday we hauled in a respectable bunch of walleyes and perch. Fortunately the Jeep started and we headed back to land. Daylight was almost gone. The only problem was the wind and blowing snow had obliterated our tracks. Jon, who was driving, became temporarily disoriented. He thought his GPS was out of whack. He wanted to head in the direction his brain told him was correct.

I took over the direction-finding. I must say, one of the best God-given gifts I have is a darn good sense of direction. I brought up an incident a few years ago when Jon and I toured the Cave of the Winds in western South Dakota. We descended several hundred feet under the surface of the earth and twisted and turned through the cave's convoluted passageways. When we reached a large underground room, the guide turned off all the lights except his flashlight. He asked if anyone knew which direction was north. I confidently raised my hand and correctly pointed northward, to the amazement of everyone on the tour. That convinced Jon to head in the direction I pointed. A few minutes later we sighted familiar landmarks we had remembered and the GPS confirmed our location.

Thirty-five minutes later we pulled up to Pete's Bar and Grill, overcrowded again. I proudly carried a half-bucket of fish with me to display to Pete and anyone else that cared to check them out.

"Is there anyone in here who isn't too drunk to clean the fish?" I asked Pete.

"Nope," Pete responded.

"Is there anyone in here who is drunk enough to clean the fish?" I asked.

"Nope," Pete remonstrated again. Handing me a knife Pete pointed to the kitchen and said, "Go at it."

Apparently a good many of the bar patrons were listening to this exchange, for I got more pats on the back as I walked toward the kitchen. Perhaps the encouragement was also meant to win a free supper, but whatever, it felt good that others were delighted in our ice-fishing success. Several other warmly dressed fishermen volunteered that they had caught only one or two fish and that most were too small to keep. The pats on the back made the fish cleaning go quickly. It didn't matter that I was posted next to the boiling hot pot of oil in which our fish would be cooked on my left side and the dishwasher on my right. No Minnesota Health Department official would dare challenge Pete about whether his sanitary practices were acceptable. The townspeople wouldn't allow such sacrilege. As I filleted the walleyes and perch the dishwasher kid told me to make sure I saved the guts and skins for the cats outside.

Jon and I sat down in the only vacant booth in the establishment. It didn't make any difference that our companion on the other side was Matty, stretched out across the seat. The beer-battered walleye tasted just the way they should. In fact all of life was the way it should be at the moment—a slice of American Pie!

Priceless

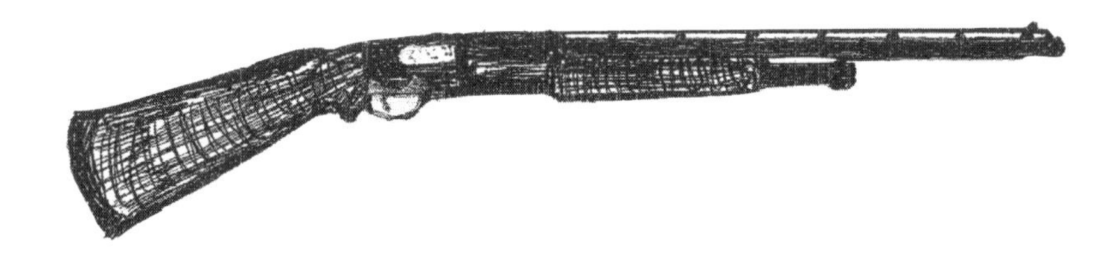

Saturday after Thanksgiving Day, 2006. Marilyn and I had a houseful of guests: Shelby and her husband, Shale, from Salt Lake City, Jon and his girlfriend from Des Moines, Marilyn's brother and sister-in-law from Milwaukee, and Shale's sister and brother-in-law from Colorado. Oh, I almost forgot Nugget. The guys and Nugget of course, all wanted to go pheasant hunting while all the ladies wanted to go shopping.

Unfortunately, instead of hunting I had to spend most of the morning at the local hospital taking care of an elderly patient who was hospitalized on Thanksgiving Day after accidentally overdosing on her medications. I got home just in time to find the ladies had gone shopping. The guys who had been afield drove up to the house as I started making lunch. They had knocked down some nice birds and Nugget had amply demonstrated her pointing and retrieving skills. Following lunch and after we cleaned up the kitchen, it was time to hunt pheasants again. Nugget was ready even though she had hunted hard for five days already.

We rode in Don's four-door Ram Charger and Jon's Jeep to the Indian grass field a half mile from home. I raised prairie grass for a seed company. The Indian grass seed had been harvested a month earlier. The combine had traveled through the field with the head set approximately four feet from the ground so it could clip the seed heads and leave the rest of the plant standing. The Indian grass stubble furnished a benevolent habitat for pheasants, deer, and lots of other animals and birds. Beaver, muskrats, and mink lived in the creek that bisected the field.

When the hunting party finished walking the east side of the creek, Don and Jon agreed to take the truck and Jeep around the road to the other side of the creek. Nugget and three of us other fellows crossed the creek on a beaver dam and continued hunting the west side. We stirred up several birds but all the roosters flushed early and were too distant to shoot at. As we finished hunting the west side of the creek, Don's Dodge Ram truck hightailed toward us through the field. When Don pulled close, he yelled through the open driver's side window, "Jon upset his Jeep in the ditch. He's okay but his truck is lying on its side."

Everybody piled into Don's truck and drove to the scene of the accident. Jon was standing next to his white Jeep, hands on his hips and looking disgusted. He explained that he didn't see the ditch in the tall grass and his Jeep tipped over when he drove too far off the path. It lay on the passenger side at the bottom of the ditch.

Everyone offered various theories about how to get the truck out of the ditch. The guys briefly tried lifting the Jeep onto its wheels but it was far too heavy to budge. Someone suggested calling a wrecker but that would be too expensive. Another person recommended using Don's 4-wheel drive Dodge Ram to pull the vehicle out of the ditch. I suggested that I could get a tractor from our farmstead. Everyone agreed that using the tractor might be the best course of action. Besides, no one had a long enough tow chain to hook the Jeep onto Don's truck. Don drove me home to fetch the tractor.

Ten minutes later I arrived with my old International 656 and a heavy-duty chain. I backed the tractor as close to the Jeep as possible and the guys hooked the chain to the front axle. Carefully creeping ahead, I pulled the vehicle onto its four wheels and slowly dragged the Jeep out of the ditch. As the truck reached the edge of the ditch, Jon jumped in through the door on the driver's side and steered it over the bank. When the truck was once again standing on level ground, the fellows inspected it thoroughly. Nothing seemed broken, not even the rear view mirror on the passenger side. A little mud and grass smeared the passenger side of the vehicle and the right front fender next to the front door was bent inward. The right front passenger door wouldn't open. What looked like oil had splashed onto the hood while the vehicle had been lying on its side.

Everyone had a theory about how to fix the Jeep so the door would open. Someone suggested that Jon shouldn't use the door on that side of the vehicle. Jon didn't think his girlfriend would like getting in on the driver's side and crawling over the transmission case to get to the passenger side, so that idea was dismissed. Someone else suggested taking the Jeep to an auto repair shop before the women arrived home from their shopping excursion but everyone decided this would cost too much. I suggested that Jon drive the vehicle to our farm shop.

The hunters reconvened by the front door of the farm shop. When I arrived back home with the tractor, they were still commiserating about repair options. While the guys shared advice, I fetched a heavy duty screwdriver and a rag.

I explained how I would pry the fender into place but everyone agreed that my intended repairs wouldn't work. Nonetheless, I wrapped the screwdriver in a rag and gently pried the fender outward. The bent fender popped into place. Everyone cheered.

Then I pried the edge of the door that had been jammed inward back into place. Everyone cheered again.

With much backslapping and ballyhooing, everyone congratulated me and commented to Jon how much money we had saved him by not taking the vehicle to a repair shop. I wiped off the engine and hood. Discerning that the *"oil"* actually was transmission fluid, I fetched a spare container of transmission fluid that I kept in my Jeep. After bringing the transmission fluid level up to *"full"* on the dipstick, everyone cheered once again.

The men cajoled Jon about how much his Dad's repairs would cost. Jon volunteered that of the three rollovers in which his vehicles had been damaged, this was by far the least expensive repair job. Then he added that his girlfriend, his mother, and none of the women in the household needed to know about today's event. Everyone cheered in agreement.

I spoke up quietly, "The cost of the towing, repair work, and transmission fluid...$5.00; the cost for not telling the ladies...priceless!"

Tachycardia

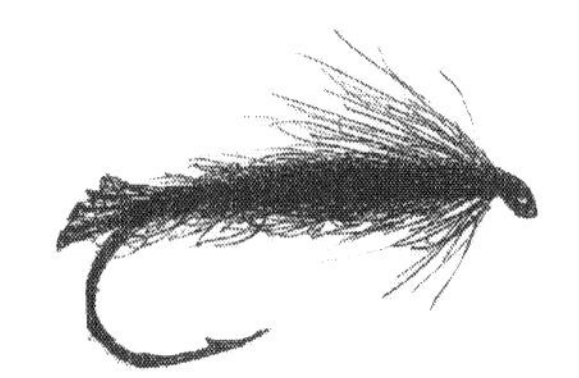

Sunday, May 18, dawned clear and warmer than the Mays I was accustomed to when I was a kid growing up in western Iowa. Good corn growing weather. Most of the corn in the neighborhood had already popped through the ground and was in the three-leaf stage. Even some soybeans had emerged, although planted only ten days ago. I could discern the neatly patterned rows while driving the country roads of western Iowa. *"It would be a good day to go fishing,"* I thought.

I knew just the place to go fishing, a pond eighteen miles away owned by my secretary and her husband. Nicely secluded, the pond lay in a pasture surrounded by steep rolling hills and swatches of timber. I had permission to drive through the chained gate and down the dirt path to the pond. The pond held some of the biggest crappie, bass, perch, and snapping turtles I had come across in Iowa or anyplace else, for that matter.

The snapping turtles made fly-fishing by float tube interesting because some of the big brutes would snatch fish tethered to the tube and occasionally sneakily nip on your fins if you didn't move your feet steadily. One time I had to fight off a snapping turtle, big as a bushel basket, for fully five minutes after it had seized my fish bag containing several bass and bluegills. I managed to kick the snapping turtle hard enough to make it release its grip on my fish bag but I had to poke at the turtle with the end of my fly rod for several minutes as it followed me around the pond until it finally gave up the fight. Usually if I tapped a turtle on the head with the tip of the fly rod, it would retreat. I learned to keep my bag of fish on the apron of my float tube so their smell wouldn't attract these amphibian submarines patrolling the waters. Particularly aggressive beasts would seize goslings, ducklings, and even nearly grown Canada Geese. Some cattlemen suspected the turtles of slashing cows' legs, teats, and udders when they waded into the pond for a drink to cool down. Cattle seldom basked in ponds that contained big turtles, even on hot summer days. I didn't mind though that there were big turtles in the pond where I was fishing because the fish in the pond were also big and numerous.

By 9:30 AM I parked my Jeep a few yards from the south shore of the pond. Twenty minutes later I was casting the fly I almost always use when going after members of the sunfish and perch families, a size six little black fly I had invented 30 years ago. Getting into the water had been a chore because the cattle in the pasture had churned up the soft mud into two-foot deep muck that extended 35 feet from the shoreline when they waded into the pond for a drink. I was glad to finally be in the cool clean water, which would wash off most of the muck. But, I didn't look forward to going through the same crap when the time came to leave the pond.

After about ten casts, an eleven-inch crappie gobbled the little black fly and soon I bagged my first fish. Crappie this big put up a nice fight and one often could not tell initially if it was a crappie or bass that had seized the fly. I was pleased with big crappie like these because they are particularly good eating and I could pass along some of the filleted bounty to my secretary. The only problem was that big fish also raised a big commotion in the water. Loud splashes and desperate underwater struggles by fish signaled marauding turtles that they might have an opportunity for dinner. I spotted a problem predator, a magnificent, nearly black beast with a head as large as my hand. I have farmer's hands, which are bigger than the hands of most men my age because cold weather, lariat and twine burns, and banging them around on farm equipment have made my fingers swell to the point that picking the guitar has become difficult. Twice the big brute started to swim toward me as I hooked frisky crappies within 20 yards of the knothead. I retreated from the aggressive turtle's territory to avoid combat.

I was feeling pretty good about things. By noon I had captured a dozen crappies, all as big as a salad plate, three bass with mouths as large as my fist, and two fifteen-inch perch. I had smoked two cigars, drunk my can of Coca-Cola, and was getting sweaty under the hot sun. Ready to quit fishing, I paddled toward my Jeep parked on the south side of the pond. When I reached the edge of the muck and could no longer paddle, I detached the float tube harness, stood up and pulled the float tube and fly rod with my left hand and dragged the heavy fish bag with my right hand through the crud. I would have to wash my waders, fins, and all my fishing equipment with a garden hose when I got home.

I struggled to yank each foot from the thick suctioning mud. It was easier to back toward the shore because then my fins created less suction but I had to take smaller steps. Once I fell backwards into the muck. I had to roll onto my knees in order to get myself upright. Huffing and puffing, I finally dragged my gear, the fish bag and myself onto dry ground. I rested on my hands and knees a bit to catch my breath. Several Angus cows and calves watched with considerable curiosity from shady spots under the bur oak trees a couple hundred feet to my left. I could feel my heart beating rapidly, at least 150 shallow ticks per minute. I wondered, *"Am I having a tachycardia attack?"*

Throughout my adult life I had experienced periodic tachycardia attacks, sometimes a single episode every few years to as many as three or four annually. When I experienced my first episode at twenty years of age, I became alarmed and frightened. I didn't know what was happening to me. Knowledge gained from my college Abnormal Psychology course suggested that perhaps this was a panic attack. I studied how to manage panic attacks: breathe slowly, deeply, and try to think of calming events. I discussed these episodes with my family doctor at the University of Virginia Medical School prior to exchanging sedentary academic life as a professor at the University of Virginia for robust physical work farming in Iowa. The precautionary stress electrocardiogram was normal. In fact, my doctor said I had remarkable endurance.

So, I didn't worry about infrequent and unpredictable tachycardia episodes. Sometimes they occurred while I was resting, other times they occurred while hoeing the garden, or stacking hay bales. One episode occurred while I was smoking a cigar, so I was tempted to give up one of my favorite outdoor recreational pursuits, but after I analyzed the circumstances surrounding the tachycardia episodes, I discounted use of tobacco and any other precipitants as having a direct cause. These consternating episodes occurred without rhyme or reason. Since all medical tests came up with inconclusive results, I didn't let the bouts of tachycardia bother me. Usually the rapid shallow heart rate cleared up after a few minutes.

Some years ago at a continuing education conference, I heard the workshop presenter report about how his patient, a farmer who had episodes of irregular heart-beat, defibrillated himself by grasping the positive pole of a tractor battery. That gave me an idea. Several months later while checking cattle with the Honda ATV, I noticed my heart beating quickly and without substance. The electric fence around the pasture was just 50 feet away. Quickly I motored up to the electric fence, got off the ATV and grabbed the wire with my bare hand. Bingo, when the shock surged through my body, the tachycardia ceased immediately. I was elated. I had discovered my own defibrillator.

The only problem with defibrillators is that sometimes they aren't around when you need them. But, I had learned how to administer a sharp blow to a calf's chest to restart the calf's heart when it had stopped during birthing. Perhaps the sharp blow caused the release of adrenaline. The next time I detected an episode of tachycardia and when an electric fence was not nearby, I hit myself on the chest under my left breast almost as hard as I could. Lo and behold, the tachycardia stopped! I had my own portable defibrillator.

I didn't report all these medical experiments to my wife, Marilyn, but I discussed them with my internist when I reached 50 years of age. I could stack hay bales all day long, set fence posts, or scoop grain for hours on end—few peers could match my capacity. When I told my internist of my tachycardia circumstances, the internist conducted numerous additional tests. I even wore a Holter monitor for several days but the results always came out normal.

Now, as I bent over on my hands and knees in muddy waders with a bag of big flopping fish beside me and my dirty fishing equipment lying ahead of me, I wasn't too worried. I stood upright, hit myself solidly on the left chest with my right fist as hard as I could three times. My hand hurt from the blows and when I unbuttoned my short-sleeved work shirt, there was a red mark under my left areola from these self-administered punches. My heart continued to flutter and I gasped for breath. Nothing hurt though, so I dragged all my equipment to the Jeep and unlaced my flippers and wading shoes. This minuscule activity made me even more winded. I had to lean over into the back of my Jeep and rest on my elbows for a full two minutes. I hit myself on the chest five more times as hard as I could stand it. Nothing happened!

"I haven't been like this before," I thought to myself. *"I wonder what's going on. I probably better stow my equipment, take my time, and watch to see if it clears up."*

I stood up and immediately became light headed. *"Oh no,"* I thought. *"Bad sign, I wonder if I'm having a heart attack. But, I don't have any pain."* I sat on the rear bumper and put my head between my knees. I uttered a brief prayer, *"Oh God, help me out here. Tell me what's going on."* No response, no change.

I raised my head from between my knees and my arms felt incredibly heavy. I stood and my legs felt as if they had heavy weights attached to them also.

"This isn't good," I thought. *"I think it's a heart attack but I don't know for sure. Maybe I should call 9-1-1 on my cell phone."*

Deliberating on this for a few seconds, I rejected the idea and silently prayed for advice for the next couple of minutes as I eased out of my waders, pulled on my jeans and shoes and lethargically stored all the fishing gear in my Jeep. Because I was so light headed and out of breath, these few activities consumed 10-12 minutes. Then I noticed my fish bag bulging with fish in the grass a few feet away. Grabbing an empty five-gallon bucket from the back of my vehicle, I stepped to the shore of the pond and allowed a couple inches of relatively clean water to cover the bottom of the bucket. Laboriously I picked up the fish bag and put it in the bucket, put the lid on and set it in the back of the Jeep. I was worn out! I thought to myself, *"Maybe I had better call 9-1-1. No, they'll never find me out here. I'll just drive home as carefully as I can and if I get too lightheaded, I'll pull off onto the roadside or into a lane or field entrance and call 9-1-1.*"

Taking much deliberate time to move slowly and to rest in between movements, I finished checking the parking area to make sure I had stored all my equipment and crawled into my Jeep. Thankfully, the engine started easily. As I slowly wound my way up the cattle path to the entrance into the pasture, I remembered that I had promised Marilyn that I would look for morel mushrooms in a wooded pasture two miles west. Last year Marilyn, our son, Jon, and I had found over five pounds of mushrooms there. With the warm moist weather of the past ten days, probably there would be a lot of morels ready to harvest. When I reached the metal pipe gate at the county road entrance, I had to make a decision: go home or hunt mushrooms first?

Stopping the Jeep and opening the gate made my limbs feel tired and heavy again. I was sweating profusely and beginning to feel clammy. I noticed my heart was still beating out of control and not accomplishing much movement of blood. "*Well, that takes care of mushroom hunting,*" I commented to myself as I chained the metal gate shut and crawled back into my Jeep. I began to head 18 miles homeward.

All the way home, I prayed, "*What does this mean? What do You want from me?*"

I was glad I had taken my usual morning aspirin. If this was a heart attack, the aspirin would help keep my blood thin and easy to circulate. I had no pain. Two miles from home I hit myself in the chest two more times with my right hand. Still no improvement!

When I reached home, I uttered a brief prayer of thanksgiving and parked my vehicle near the water faucet on the outside wall of the house. Taking my time, I pulled the muddy float tube, waders, fins, and the five-gallon bucket half full of fish onto the lawn. Turning on the water spigot, I washed everything clean, even my vehicle that was dusty. Then I grabbed the bucket of fish and carried it into the back entryway. I was still breathing heavily, sweating and pale complected. Kicking off my shoes and leaving the bucket of fish inside the back entrance, I stepped into the living room. Hearing me, Marilyn called out, "What are you doing home so early?"

"I got all the fish I wanted but I don't feel good," I said. "I think I'm having a heart attack."

Marilyn shrieked, "I always knew this would happen! When did this happen? Are you okay? How come you didn't call me? We have to get you the emergency room. Did you take your aspirin this morning? Who's on call at the hospital? Should I call 9-1-1? Are you having pain? What does it feel like? You look clammy. You need to lie down on the couch. I'll get you some water and aspirin."

"I'm having tachycardia and it won't go away like it usually does when I hit myself on the chest," I responded calmly. "Yes, I took my aspirin this morning."

Fumbling through the medicine cabinet, Marilyn was unable to find the aspirin bottle. I got up and found the aspirin bottle on the top shelf, in its usual place, but too high for Marilyn to easily see. After all, she is a foot shorter than me.

"Probably we should go to the emergency room." I volunteered. "I left my Jeep outside and I washed off all my fishing equipment. I left the fish in the back entryway."

"How many did you catch?" Marilyn asked.

"I got about a dozen great big crappie, three bass about three pounds apiece, and two nice perch," I reported. "Make sure you don't throw them away. I'll clean them when we get home," I instructed. "Freeze them or have Jon clean them when he comes home."

"We better go to the hospital right away," Marilyn pronounced.

"I need to take a shower first," I rejoined. "The staff at the emergency room won't like me if I come in looking and smelling like this."

I made my way to the bathroom, showered and donned clean clothes. I grabbed my toothbrush, some fishing magazines and said "Let's go," to Marilyn, who was searching for her car keys.

Two hours and $2,100 later, the emergency room physician announced to Marilyn and me, "It doesn't look like you had a heart attack because your troponin level is normal but I'm concerned that you were short of breath, clammy, and your arms and legs felt heavy. I've called for the ambulance to take you to the cardiac unit in Omaha."

"How about if Marilyn and I drive to Omaha ourselves?" I asked.

"Oh no, we could never allow anything like that!" the doctor retorted.

Two hours and another $2,000 later, I was in the cardiac unit at University Hospital in Omaha. Marilyn had called our son in Des Moines, and our daughter, an internist and rheumatology fellow in Salt Lake City. Shelby was particularly helpful discerning medical options and Jon was understandably worried that his father and fishing buddy could be in dangerous straits. I confessed to everyone how I managed my bouts of tachycardia for years by grabbing electric fences and hitting myself in the chest, and that I knew this episode was more than just tachycardia.

Two days, two stents, and $100,000 later, the various specialists explained that my left anterior descending artery in my heart was completely blocked and that probably my tachycardia bouts were indications of advancing blockage over many years. But, collateral circulation had developed to the point that I had no heart damage. I could go fishing and hunting all I wanted.

As we were leaving the University Hospital, I quietly asked Marilyn, "What did you do with the fish I caught last Sunday?"

"Nothing, we gave them to the cats. I won't clean fish, you know. And Jon came here to see you instead of cleaning the fish," Marilyn responded.

"Guess I'll have to go fishing again soon," I admonished.

The Behavior of Farm People

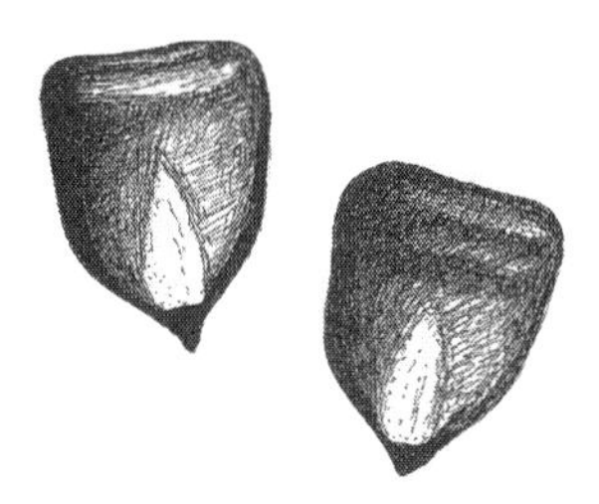

As the second semester of college resumed at the University of Virginia following Christmas break in mid-January 1979, I announced to my dozen or so graduate students and research assistants that I planned to resign from the Psychology Department faculty at the end of the semester to move to a family farm in western Iowa and initiate a clinical psychology practice there. Several snickers arose, followed by muffled laughter, and then stiff silence. A third-year graduate student whose master's thesis I was supervising asked, "Why are you leaving the University of Virginia to work with farmers?"

"Somebody has to take care of the mental health of farm people," I answered, but the words didn't seem like they were mine. A deep inner force I couldn't explain, seemed to take charge of my speaking.

For the next several minutes I found myself proclaiming that family farmers in the United States are an endangered species and a national treasure to be preserved. The students, like so many people, had come to take food for granted and had lost connections with the sources of their nourishment. I felt deep inner urges to defend agriculture and the people who work the land. I was exhilarated, resolute, and excited to be embarking on an adventure with a course as yet unknown.

Twenty weeks later I was sitting on a tractor seat and planting corn with my father clinging to the fender on my left side. "Keep an eye on how the planter wants to slide down the hillside so your rows aren't as straight as they could be," Dad cautioned. "I'll make another round or two with you but then you are on your own," he chippered and smiled with a look of pride in his eyes.

I was the second son in the family to come back to the farm. Dad and Mom had not counted on us returning to farm. They figured Marilyn and I would continue as nursing and psychology professors and our children, Shelby and Jon, would grow up on our three acres in Virginia near Meriwether Lewis' family home. They withheld judgment when we announced our intentions during our Christmas visit, but commented positively how much thought we had given over the past five years to raising our family on a farm. Marilyn added the crowning observation, that we had a huge garden on our acreage and, "Mike is always going outside the house to check on the weather." She commented how she had wanted to marry a farm boy, but she didn't expect she would someday live and work on a real Iowa farm.

Coming to America for Marilyn's and my ancestors was risky. Few knew before their arrival where they would find their parcels on which to till the soil and raise animals for their livelihoods. One of my great grandfathers was a shoestring peddler and tinker until he had earned enough money at age 29 to purchase a team of horses and marry a 19-year-old immigrant, a German Catholic woman with a similar passion for farming. Farm they did, and passed along the heritage to their 14 children, and up to this point, five more generations! Most farmers are like that. They view farming as a noble calling, a vocation akin to becoming a priest, nun, teacher, or physician. Producing necessities for life—the food, fiber, and now increasingly the renewable energy we need to sustain life—satisfies farmers' deepest urges to feel appreciated, essential, and to give.

To farmers the land means everything. Ownership of a family farm is the triumphant result of the struggles of multiple generations of immigrants to America. Losing the family farm is the ultimate loss—bringing shame to those who've let down their predecessors while dashing hope for successors. Farming has always been one of the most stressful and dangerous occupations, in the top three along with fishing and forestry, which certainly have much in common with agriculture. Much of the stress traces to having little control over many of the factors that affect production, like the weather, disease outbreaks, government policy, and supply/demand conditions. The emotional well-being of family farmers is intimately entwined with this way of life.

But often farmers can't—or won't—talk about their love affair with farming. When the opportunity to continue farming is threatened, they keep it to themselves. Their intense personal struggles to maintain self respect in the face of overwhelming losses usually are private but the repercussions affect the whole family. One Iowa farmer said it poignantly in a suicide note to his family. His wife shared his note with an Iowa newspaper after her husband hung himself on a Sunday morning while she and their two kids attended church. "The only thing I will regret is leaving the children and you. This farming has brought me a lot of memories, some happy, but most of all grief. The grief has finally won out—the low prices, bills piling up, just everything. The kids deserve better and so do you. I just don't know how to do it. This is all I know and it's just not good enough anymore."

When I first initiated my psychology consulting practice, I couldn't find the words to express what I was feeling about the emotional underpinnings of farmers and ranchers. Gradually I came to realize that "agriculture" has two main stems, "agri" and "culture," and the latter is the more important of the two. Understanding the culture of farm people is critical to being an effective psychological counselor to this complicated set of people.

I thought farmers would be reluctant to seek help for mental health problems even when they need it. The words "mental health" generate images of unconscious impulses that can be controlled only if we exert enough willpower or take enough psychotropic medication. To many farmers, mental health and substance abuse problems are signs of weakness that can be discussed only with a pastor, family doctor, or trusted family members. But, did the farmers ever seek me out! At first I wasn't sure why farm people opened up to me. A cattleman who was depressed and had to sell all his cows to pay off foreclosed loans explained it better than I could when he said, "You are a farmer like us ... you understand." Within two years after opening my psychology practice, I was both a full-time farmer and psychologist. I approached the nearest community mental health center in a neighboring town to see if they would hire me so I could have backup. The center hired me.

Being a farmer helped me become a better psychologist and vice versa. The farm people and I could speak the same language. Implicitly, we shared the same culture. They helped put into words the motivations that I felt when I gave my resignation speech to my University of Virginia colleagues. Many would make their first call to me on my home phone. Often they were scared and ashamed. They wanted to initiate contact in a way that seemed more comfortable and less formal than calling the office. It was tough on my family and me because farm people would call at all hours of the day or night. But it was—and still is—affirming when a grateful client says, "You helped me."

In the past, younger farmers with college education were more open to getting professional mental health assistance than older farmers, but this is gradually changing. Now most farm owners of all ages look for whatever will help them remain viable. We are not as effective reaching ethnic minorities involved with agriculture, and traditional groups such as Mennonites, but we need to try to understand their cultural backgrounds.

We learned during the Farm Crisis of the 1980s that farmers also will call telephone hotlines if they are manned by responders who understand agriculture and if confidentiality and anonymity are guaranteed. The Farm Crisis taught us that losing the family farm is worse than the death of a family member. Adjustment to the loss of a family farm follows a series of stages akin to Kübler-Ross' stages of grief but the trauma is more severe and there are fewer rituals to help people handle their losses.

It is interesting how farm culture dwindles out of people, as it had from my Virginia cohorts, when they leave their rural roots to work in towns and cities. Usually farm culture washes out after three successive generations have been removed from working the land. I noticed, though, that farm culture can be rekindled in a single generation. City folks who marry farm people or who move to rural areas to begin their own farming operations acquire the yen to work the soil in just a few years and pass along to their children their hardy work ethic and the spiritual commitment to make the land produce. This rapid reversion to farm culture probably taps into strands of genetic memory that lie dormant until agricultural activities stimulate the emergence of a wealth of survival skills that are included in their DNA. Domestic pigs and dogs will revert back to being wild in a single generation when left untended by human caretakers. Boars' tusks elongate, their hair thickens, and their behavior becomes wary. Dogs form packs and pool their resources to capture prey and protect their puppies. Humans are sophisticated animals.

What is it that stimulates farmers deep abiding attachment to the land? Is it sunlight? Or perhaps getting dirt under our fingernails and absorbing the molecules of dust into the molecular structure of body cells? Maybe it's the smell of fresh mown hay or the not-so-disagreeable pungency of cattle's breath as they crowd up to the bunks when the feeder wagon delivers silage and grain. Or maybe it's observing dots of corn emerge from May soil into discernable patterned rows of green plants. It's all of these and more.

I proposed a construct, the agrarian imperative, as the best explanation why people engage in agriculture. I offered several lines of evidence as validation of the agrarian imperative. Historical evidence suggests that domesticating animals and cultivating land to produce food, fiber, and shelter allowed humans to survive lean times and to proliferate faster than hunter-gatherers. Agriculture yielded survival advantages for the human species. Genetic and anthropological evidence is accruing which suggests that acquiring territories of land and sea to produce these necessities has an inherited basis which is encoded into our genetic material. Feedback from the environment influences and modifies the genetic memory pertaining to the agrarian imperative. Psychological evidence, particularly personality research, suggests behavioral traits that are characteristic of persons who engage in agriculture: great capacity to cope with adversity, conscientiousness, risk-taking, and self-reliance. Inability to farm successfully, however, is associated with a profound sense of failure and an increased probability of suicide.

The Farm Crisis took a personal toll. In the early 1980s we bought more land—my grandmother's farm and 80 acres my mother sold to pay for the care of Larry, my disabled younger brother. I employed a hired hand who worked nearly full-time in our farm operation as we built a purebred Simmental cattle operation to raise breeding stock and began the transition of our conventional farm operation to become an organic farm enterprise. Mostly by ourselves, we built two large machine storage sheds, a shop, two big metal grain bins, two cattle barns, and several sheds for the calves, and for the turkeys and broilers cared for mostly by Shelby and Jon. By now Marilyn was reminding me that the livestock had better facilities to live in than our family. Our 1870s four-square farm house was cold and drafty in the winter and too hot in the summer. To prove how uncomfortable our uninsulated clapboard structure could be, one cold winter night Marilyn set a bucket of water under the north window of the upstairs bedroom we shared. By morning a crust of ice covered the bucket. Marilyn boldly threatened to stuff it down the back of my shirt if I refused to promise that the next building would be a new house. "You remind me of the wife in a Hamlin Garland story," I teased her. "She set up house in the new barn while her husband was away from home purchasing more cows."

"Just buy a few more cows and see what happens," Marilyn laughed. That night after supper I began drafting house plans.

By the end of the decade I had left the neighboring mental health center and took on the challenge of founding Prairie Rose Mental Health Center in Harlan, Iowa, in response to a request from the Shelby County Board of Supervisors and the hospital trustees. I was working practically day and night. The mild ADHD that I grew up with served me well, for I didn't need much sleep—four or five hours was enough. Work caught up with me on July 24, 1990, when in a moment of haste and poor judgment I stuck my right foot into the auger that unloaded the hopper of the grain combine. In an instant my foot was dragged under the auger blade and it severed all but my big toe. Instinctively, I yanked my foot with such force that I tore the heavy lug sole off the bottom of my work shoe and jumped from the hopper. But for a moment I pondered the blood spurting from my right foot and thought, *"a few stitches and I should be okay."* Horror swept over me in the next instant as I realized that the injury was worse than I thought initially and I was in for some downtime. Then a question popped into my mind, *"Oh, God what do you want from me?"* All of this in less than a second-and-a-half.

It took two weeks in the hospital, lots of prayers, tears and support from my wife and children, my hired hand, and neighbors to figure out the answer to this paramount question. Instantaneously I had become vulnerable. Gradually, I realized that my own behaviors were screwed up. I was trying to do too much and partially for the wrong reasons. A higher force was telling me that my behaviors and my motives were less than completely healthy. I shouldn't work so hard, or maybe work smarter. I should concentrate more on helping others and focus less on material gains. Over the next three years we sold Grandma's and Larry's farms. I gave up two Board positions and we began as a family to camp, fish, play more, and nurture our relationships. I learned to put work aside for a day here or a few hours there to go fishing. It must have looked odd to the neighbors when I let the combine set in a partially harvested soybean field on a sunny October afternoon while I drove my pickup truck loaded with a canoe and fishing gear past them on the country roads. I had come to realize that taking time to recreate was an important investment in me. Just as a feed ration has ingredients that can be varied to maximize growth of beef cattle, behavior can be rationed to maximize well-being. We are in charge of our behavior ration, how much and how hard we work, sleep, recreate, pray, laugh, talk, and so forth.

As I finished graduate school in the mid-1970's my approach in clinical work was to help clients "work through" issues. My job was to conduct therapy that led to a cure. Gradually the field of clinical psychology has evolved to emphasize behavior management and I evolved along with it. Behavior management entails coaching the client about how to govern himself. It places the client in charge of making choices instead of requiring the therapist to be responsible for changing the client. For example, when a farmer is depressed, in the past I would help the client work through what was making him depressed, such as conflict with his wife and inadequate finances. Now I help the client take charge of those things he can be responsible for. He may not be able to change his conflict or finances but he can exercise to produce his own serotonin and he can insure that he gets adequate sleep and avoids exposures to insecticides because all these things contribute to depression. In short, the client can manage the symptoms of depression. When he is less depressed, he can think better, work more effectively, and is easier to be around. In turn, the conflict with his spouse and his financial situation both improve.

My advice is not necessarily to work less hard, but to work smarter. For example, taking breaks improves concentration and output. Adequate sleep improves thinking and concentration. Did you know that a person with a sleep debt of just 10 hours over a two-week period is as unsafe as a person with a .08 blood alcohol level?

To illustrate further, a person who is prone to Seasonal Affective Disorder can enhance the likelihood of avoiding depressive lows by replacing incandescent light bulbs in the farm house, shop, and barns that the farmer frequents with fluorescent tubes with the same light wavelength as sunlight. Conversely, manicky highs can be curtailed by wearing a wide brimmed cap or hat that reduces the amount of sunlight that enters the eyes and which triggers the release of pineal gland hormones that excite the brain when days are long. Even the ingestion of Lithium

Carbonate, a chemical regulator of this mood disorder, is a behavior that can be governed or ignored. We can engage in the behaviors that counter the symptoms that accompany maladjustment—like talking about tender feelings when we most want to hide our emotional pain.

In 1998 I resigned from Prairie Rose Mental Health Center to devote all my professional time to improving the behavioral healthcare of the agricultural population. A new field, agricultural behavioral health, was emerging. I joined the College of Public Health staff of the University of Iowa and began to travel, lecture, and publish widely. AgriWellness, Inc., a multi-state nonprofit organization with a home base in Harlan, took shape. We formed a network of seven upper Midwestern states, each with a farmer-friendly telephone helpline than can link callers with a variety of necessary supports. Now we are proposing a national network of behavioral health supports to help agricultural producers throughout the country cope with stresses of all types

Researchers have found that less than half of all Americans have the good fortune of pursuing a vocation they like during their lives. I have had the opportunity to respond to two primal callings, agriculture and behavioral healthcare, which now are merging into a single field, while also pursuing ample rations of wonderful recreation: fishing, hunting, and writing. Agricultural behavioral health is now being taught to aspiring physicians, nurses, and behavioral health professionals such as psychologists in several universities and many workshops around the country. Journal articles on the subject are accruing. Many magazine articles in farm publications and radio (e.g., National Public Radio) and television (e.g., CNN, *National Geographic Today*) programs are popularizing the term.

The future of agriculture in the United States and many other countries is moving toward two types of farming. One is the industrial model that emphasizes specialization of production, economy of scale, and the careful management—even the exploitation—of all resources, including the people in farming. The other is the sustainable model that emphasizes conservation of resources and diversity of production (e.g., crops and livestock), often by organic methods, and which can be described as, "farming as a way of life." It is interesting that the children of organic farmers are three times more likely to enter farming than the children of industrial model operators. But this is understandable because if our hearts aren't into it, we don't find meaning. For farm people who do this as our life's work, farming is a sacred act. Understanding the behavior of farm people is my life's work.

Photo by American Images, Inc.

The Rosmann farmstead in western Iowa. The village of Westphalia is visible in the background.

Photo by Mike Rosmann.

The author's son, Jon, in the midst of a cherished moment of "Excellent Joy."

Photo by Mark Jacobs.

Larry Rosmann, the author's brother and inspiration for *Excellent Joy*.

Photo by Steve Gibbons.

The "Little Black Fly."

Photo by Jon Rosmann.

Mike enjoying peace in the field.

Photo by Jon Rosmann.

Nugget doing what she loves.

Photo by Mike Rosmann.

The land, livestock, and the family farm mean everything to farmers and are the reason why the author has focused his life's work on agricultural behavioral health.

Photographer unknown.

Legendary fly fishing teacher and the author's father-in-law, Walt.

Photo by Mark Jacobs.

Mike Rosmann ready to take to the stream.

Michael R. Rosmann is a psychologist and farmer whose life's work involves improving the behavioral healthcare of the agricultural population. He seeks to advance regional and global food production policy which enhances the behavioral and economic welfare of food producers, maintains stewardship of the land and other resources used in food production and protects the safety of food for consumers. In an era of increasing tension due to bioterrorist threats and shifts in the agribusiness climate, he is a voice for the agricultural population. *The New York Times* said this about him: "A fourth generation farmer as well as a clinical psychologist, he speaks the language of men and women on the verge of losing their place on the land."

Rosmann has been instrumental in developing a new specialty: agricultural behavioral health. He has appeared on ABC, CBS, CNN, and National Geographic television network programs and has been a guest on National Public Radio and the Farm Bureau Network numerous times. In addition, he has served as a keynote speaker at many state, regional, national, and international conferences. With other concerned citizens, Rosmann founded AgriWellness, Inc., a nonprofit seven-state program which promotes accessible behavioral health services for underserved and at-risk populations affected by the farm crisis and by the ongoing transitions in agriculture. He is adjunct faculty at the University of Iowa. Rosmann is the author of many scholarly and popular articles, book chapters, short stories, and an avid fly fisherman and hunter.

Ice Cube Books began publishing in 1993 to focus on how to live with the natural world and to better understand how people can best live together in the communities they share and inhabit. Since this time we've been recognized by a number of well-known writers including: Gary Snyder, Gene Logsdon, Wes Jackson, Patricia Hampl, Greg Brown, Jim Harrison, Annie Dillard, Ken Burns, Kathleen Norris, Janisse Ray, Alison Deming, Richard Rhodes, Michael Pollan, and Barry Lopez. We've published a number of well-known authors including: Mary Swander, Jim Heynen, Mary Pipher, Bill Holm, Connie Mutel, John T. Price, Carol Bly, Marvin Bell, Debra Marquart, Ted Kooser, Stephanie Mills, Bill McKibben, and Paul Gruchow. As well, we have won several publishing awards over the last eighteen years. Check out our books at our web site, join our facebook group, visit booksellers, museum shops, or any place you discover good books to discover why we strive to "hear the other side."

Ice Cube Press (est. 1993)
205 N Front Street
North Liberty, Iowa 52317-9302
steve@icecubepress.com
www.icecubepress.com

back we go
to those river tips &
tree trunks
many fluid & sun reaching
thanks to my co-horts
Fenna Marie & Laura Lee

Printed and bound by PG in the USA